Integration of High Voltage AC/DC Grids into Modern Power Systems

Integration of High Voltage AC/DC Grids into Modern Power Systems

Special Issue Editor

Fazel Mohammadi

MDPI • Basel • Beijing • Wuhan • Barcelona • Belgrade • Manchester • Tokyo • Cluj • Tianjin

Special Issue Editor
Fazel Mohammadi
University of Windsor
Canada

Editorial Office
MDPI
St. Alban-Anlage 66
4052 Basel, Switzerland

This is a reprint of articles from the Special Issue published online in the open access journal *Applied Sciences* (ISSN 2076-3417) (available at: https://www.mdpi.com/journal/applsci/special_issues/modern_power).

For citation purposes, cite each article independently as indicated on the article page online and as indicated below:

LastName, A.A.; LastName, B.B.; LastName, C.C. Article Title. *Journal Name* **Year**, *Article Number*, Page Range.

ISBN 978-3-03936-525-8 (Hbk)
ISBN 978-3-03936-526-5 (PDF)

Contents

About the Special Issue Editor

Fazel Mohammadi is the founder of the Power and Energy Systems Research Laboratory. He is a Senior Member of Institute of Electrical and Electronics Engineers (IEEE), and an active member of International Council on Large Electric Systems (CIGRE), European Power Electronics and Drives (EPE) Association, American Wind Energy Association (AWEA), and the Institution of Engineering and Technology (IET). His research interests include power and energy systems control, operations, planning and reliability, high-voltage engineering, power electronics, and smart grids.

Editorial

Integration of High Voltage AC/DC Grids into Modern Power Systems

Fazel Mohammadi

Electrical and Computer Engineering (ECE) Department, University of Windsor, Windsor, ON N9B 1K3, Canada; fazel@uwindsor.ca or fazel.mohammadi@ieee.org

Received: 22 May 2020; Accepted: 22 May 2020; Published: 26 May 2020

Abstract: The Special Issue on "Integration of High Voltage AC/DC Grids into Modern Power Systems" is published. A total of five qualified papers are published in this Special Issue. The topics of the papers are control, protection, operation, planning, and scheduling of high voltage AC/DC grids. Twenty-five researchers have participated in this Special Issue. We hope that this Special Issue is helpful for high voltage applications.

Keywords: High Voltage AC/DC Grids; Power Systems Control; Power Systems Operation; Power Systems Optimization; Power Systems Planning; Power Systems Protection

---

1. Introduction

Electric power transmission relies on AC and DC grids. The large integration of the conventional and non-conventional energy sources and power converters into power grids has resulted in a demand for High Voltage (HV), Extra-High Voltage (EHV), and Ultra-High Voltage (UHV) AC/DC transmission grids in modern power systems [1–3]. To ensure the security, adequacy, and reliable operation of power systems, practical aspects of interconnecting HV, EHV, and UHV AC/DC grids into the electric power systems, along with their economic and environmental impacts should be considered. The stability analysis for planning and operation of HV, EHV, and UHV AC/DC grids in power systems is regarded as the other key issue in modern power systems [4,5]. Moreover, interactions between power converters and the other power electronics devices (e.g., FACTS devices) installed on the network are the other aspects of power systems that must be addressed [6]. This Special Issue aims to investigate the integration of HV, EHV, and UHV AC/DC grids into modern power systems by analyzing their control, operation, protection, dynamics, planning, reliability, and security along with considering power quality improvement, market operations, power conversion, cybersecurity, supervisory and monitoring, diagnostics, and prognostics systems.

2. Integration of High Voltage AC/DC Grids into Modern Power Systems

M. J. Alvi, et al. [7], in their paper entitled "Field Optimization and Electrostatic Stress Reduction of Proposed Conductor Scheme for Pliable Gas-Insulated Transmission Lines", performs the geometric and electrostatic field optimization for Flexible Gas-Insulated Transmission Lines (FGILs) regarding stranded conductors. Also, the impact of conductor irregularity on field dispersion is investigated, and a Semiconducting Film (SCF)-coated stranded conductor is suggested as a potential candidate for FGILs. Owing to the performed optimized design, an 11 kV scaled-down model of a 132 kV FGIL is fabricated to practically investigate the electrostatic and dielectric stresses in the FGIL through an HV experimental setup.

F. Mohammadi, et al. [8], in their paper entitled "An Improved Mixed AC/DC Power Flow Algorithm in Hybrid AC/DC Grids with MT-HVDC Systems", proposes a mixed AC/DC Power Flow (PF) algorithm for hybrid AC/DC grids with Multi-Terminal High-Voltage Direct Current (MT-HVDC) systems. The proposed strategy is a fast and accurate method, which is capable of optimizing the AC/DC PF calculations. Except for the high accuracy and optimized performance, considering all operational constraints and control objectives of the integration of MT-HVDC systems into the large-scale AC grids is the other contribution of this paper. The calculated results by the mixed AC/DC PF problem can be used for the planning, scheduling, state estimation, small-signal stability analyses. The mixed AC/DC PF algorithm is applied to a five-bus AC grid with a three-bus MT-HVDC system and the modified IEEE 39-bus test system with two four-bus MT-HVDC systems (in two different areas), which are all simulated in MATLAB software. To check the performance of the mixed AC/DC PF algorithm, different cases are considered.

A. H. Shojaei, et al. [9], in their paper entitled "Multi-Objective Optimal Reactive Power Planning under Load Demand and Wind Power Generation Uncertainties Using ε-Constraint Method", attempts to address Reactive Power Planning (RPP) as a probabilistic multi-objective problem to reduce the total cost of reactive power investment, minimize the active power losses, maximize the voltage stability index, and improve the loadability factor. The generators' voltage magnitude, the transformers tap settings, and the output reactive power of the VAR sources are considered as the main control variables. To deal with the probabilistic multi-objective RPP problem, the ε-constraint technique is employed. To validate the efficiency of the proposed method, the IEEE 30-bus test system is implemented in the GAMS environment under five various conditions.

Y. Cui, et al. [10], in their paper entitled "Effect of Ionic Conductors on the Suppression of PTC and Carrier Emission of Semiconductive Composites", discusses the Positive Temperature Coefficient (PTC) effects of electrical resistivity in perovskite $La_{0.6}Sr_{0.4}CoO_3$ (LSC) particle-dispersed semiconductive composites of HVDC cables based on experimental results from Scanning Electron Microscopy (SEM), Transmission Electron Microscopy (TEM) and a semiconductive resistance test device.

T. T. Nguyen, et al. [11], in their paper entitled "Optimal Scheduling of Large-Scale Wind-Hydro-Thermal Systems with Fixed-Head Short-Term Model", implements a Modified Adaptive Selection Cuckoo Search Algorithm (MASCSA) for determining the optimal operating parameters of a hydrothermal system and a wind-hydro-thermal system, to minimize the total electricity generation cost from all available thermal power plants. The fixed-head short-term model of hydropower plants is taken into consideration. All hydraulic constraints, such as initial and final reservoir volumes, the upper limit and lower limit of reservoir volume, and water balance of reservoir, are seriously considered. The proposed MASCSA competes with the conventional Cuckoo Search Algorithm (CSA) and Snap-Drift Cuckoo Search Algorithm (SDCSA). Two test systems, (1) four hydropower plants and four thermal power plants with valve effects over one day with twenty-four one-hour subintervals, and (2) four hydropower plants, four thermal power plants, and two wind farms with the rated power of 120 MW and 80 MW over one day with twenty-four one-hour subintervals, are employed to check the validity and accuracy the proposed method and compare its performance with the mentioned CSA-based methods.

Funding: This research received no external funding.

Acknowledgments: I am thankful for the contributions of professional authors and reviewers and the excellent assistant of the editorial team of *Applied Sciences*.

Conflicts of Interest: The author declares no conflict of interest.

References

1. Mohammadi, F.; Nazri, G.-A.; Saif, M. An Improved Droop-Based Control Strategy for MT-HVDC Systems. *Electronics* **2020**, *9*, 87. [CrossRef]
2. Mohammadi, F.; Nazri, G.-A.; Saif, M. A Bidirectional Power Charging Control Strategy for Plug-in Hybrid Electric Vehicles. *Sustainability* **2019**, *11*, 4317. [CrossRef]

3. Mohammadi, F. Power Management Strategy in Multi-Terminal VSC-HVDC System. In Proceedings of the 4th National Conference on Applied Research in Electrical, Mechanical Computer and IT Engineering, Tehran, Iran, 4 October 2018.

4. Mohammadi, F.; Zheng, C. Stability Analysis of Electric Power System. In Proceedings of the 4th National Conference on Technology in Electrical and Computer Engineering, Tehran, Iran, 27 December 2018.

5. Mohammadi, F.; Nazri, G.A.; Saif, M. A Fast Fault Detection and Identification Approach in Power Distribution Systems. In Proceedings of the IEEE 5th International Conference on Power Generation Systems and Renewable Energy Technologies (PGSRET), Istanbul, Turkey, 26–27 August 2019.

6. Nguyen, T.T.; Mohammadi, F. Optimal Placement of TCSC for Congestion Management and Power Loss Reduction Using Multi-Objective Genetic Algorithm. *Sustainability* **2020**, *12*, 2813. [CrossRef]

7. Alvi, M.J.; Izhar, T.; Qaiser, A.A.; Kharal, H.S.; Safdar, A. Field Optimization and Electrostatic Stress Reduction of Proposed Conductor Scheme for Pliable Gas-Insulated Transmission Lines. *Appl. Sci.* **2019**, *9*, 2988. [CrossRef]

8. Mohammadi, F.; Nazri, G.-A.; Saif, M. An Improved Mixed AC/DC Power Flow Algorithm in Hybrid AC/DC Grids with MT-HVDC Systems. *Appl. Sci.* **2020**, *10*, 297. [CrossRef]

9. Shojaei, A.H.; Ghadimi, A.A.; Miveh, M.R.; Mohammadi, F.; Jurado, F. Multi-Objective Optimal Reactive Power Planning under Load Demand and Wind Power Generation Uncertainties Using ε-Constraint Method. *Appl. Sci.* **2020**, *10*, 2859. [CrossRef]

10. Cui, Y.; Yin, H.; Xing, Z.; Guo, X.; Zhao, S.; Wei, Y.; Li, G.; Xin, M.; Hao, C.; Lei, Q. Effect of Ionic Conductors on the Suppression of PTC and Carrier Emission of Semiconductive Composites. *Appl. Sci.* **2020**, *10*, 2915. [CrossRef]

11. Nguyen, T.T.; Pham, L.H.; Mohammadi, F.; Kien, L.C. Optimal Scheduling of Large-Scale Wind-Hydro-Thermal Systems with Fixed-Head Short-Term Model. *Appl. Sci.* **2020**, *10*, 2964. [CrossRef]

Article

Field Optimization and Electrostatic Stress Reduction of Proposed Conductor Scheme for Pliable Gas-Insulated Transmission Lines

Muhammad Junaid Alvi [1,2,*], Tahir Izhar [1], Asif Ali Qaiser [3], Hafiz Shafqat Kharal [1] and Adnan Safdar [2]

[1] Department of Electrical Engineering, University of Engineering and Technology, 54890 Lahore, Pakistan
[2] Department of Electrical Engineering, NFC Institute of Engineering and Fertilizer Research, 38090 Faisalabad, Pakistan
[3] Department of Polymer Engineering, University of Engineering and Technology, 54890 Lahore, Pakistan
* Correspondence: engr.junaidalvi@iefr.edu.pk; Tel.: +92-333-653-3929

Received: 13 June 2019; Accepted: 16 July 2019; Published: 25 July 2019

Featured Application: Flexible gas-insulated transmission lines (FGILs) are a potential candidate for the trenchless underground implementation of high-voltage transmission lines in metropolitan areas. This research highlights the necessity of field intensity minimization and field irregularity suppression for FGILs regarding stranded conductors and proposes a practicable scheme for the same. The proposed scheme will facilitate the achievement of analogous electrostatic and dielectric characteristics for FGILs as compared to conventional gas-insulated lines (GILs).

Abstract: The implementation of stranded conductors in flexible gas-insulated transmission lines (FGILs) requires field intensity minimization as well as field irregularity suppression in order to avoid dielectric breakdown. Moreover, the interdependence of enclosure and conductor sizes of FGILs regarding electrostatic aspects necessitate critical consideration of their dimensional specifications. In this research, geometric and electrostatic field optimization for FGILs regarding stranded conductors is performed. In addition, the effect of conductor irregularity on field dispersion is analyzed, and a semiconducting film (SCF)-coated stranded conductor is proposed as a potential candidate for FGILs. Considering the performed optimized design, an 11 kV scaled-down model of a 132-kV FGIL was also fabricated in order to practically analyze its electrostatic and dielectric performances regarding simple and SCF-coated stranded conductors. Simulation and experimental investigations revealed that the SCF-coated stranded conductor significantly minimized the field irregularity of the FGIL along with improving in its dielectric breakdown characteristics.

Keywords: dielectric strength; field grading; field utilization factor (FUF); gas-insulated transmission line; metropolitan; stranded conductor

---

1. Introduction

Escalating urbanization and industrialization has resulted in an increased load demand along with the necessity of higher system stability and reliability, which requires the upgrade and new installation of power transmission schemes (PTSs) [1–5]. Moreover, renewable energy integration [6,7], smart grid development [5,8], and the need of interruption-free operation in the case of faults [8,9] also require the implementation of PTSs within metropolitan areas [10,11]. Researchers have described that conventional PTSs include overhead lines (OHLs) [1,3,5,10,12], underground cables (UGCs) [7,13,14] and gas-insulated lines (GILs) [15–18]. Literature regarding the metropolitan application of PTSs mentioned that OHLs and UGCs encounter hindrances such as right of way [2,3,19], spatial proximity [9,20,21], aesthetics [19,20], system failure due to prolonged fault clearance time [8,9], corrosion [2,22], trench requirements [14,23], and electromagnetic compatibility (EMC) concerns [2,4,21,24,25]. Further, studies mentioned that conventional GILs also face impediments regarding their implementation in urban vicinities due to their metallic profile, such as their structural rigidity [15,25,26], larger bending radius and lay length [15,16,27], jointing complexities [15,17,27], corrosion protection [16,24,28], requirement of acceleration dampers [17,24,29], and trench development [11,27,30]. Thus, protruding urbanization, despite being a potential load consumer, critically curtails the implementation of conventional PTSs in metropolitan vicinities.

References [11,31–34] reveal that flexible gas-insulated lines (FGILs) comprised of a reinforced thermoplastic enclosure, stranded conductor, and polyurethane (PU) post insulator are a potential candidate for curtailing the intricacies associated with the implementation of conventional PTSs in metropolitan areas. Further, researchers [35–37] have mentioned that flexible cables and enclosures like FGILs are practicable for horizontal directional drilling (HDD)-based underground laying schemes and do not require trench development, which is highly beneficial in urban vicinities. Thus, the simplification of several issues associated with conventional PTSs like right of way, EMC concerns, trench requirement, corrosion protection, and larger land area requirement makes FGILs an appropriate scheme for the subsurface metropolitan application of high-voltage lines. However, researchers have mentioned that the contour irregularity of stranded conductors [38] is a point of concern due to its irregular field distribution [39,40], which results in poor field utilization [17,24,41,42] and augments partial discharge activity [43–45] and streamers [43,45,46]. Moreover, references [17,24,41,42] mentioned that the interdependence of enclosure and conductor sizes apropos of field utilization necessitate critical consideration regarding the dimensional specifications of FGILs in case of any variation in their field utilization. Thus, field irregularity due to stranded conductors in FGILs along with its effect upon dimensional specification needs thoughtful consideration.

Researchers [47–56] mentioned that regarding GILs, irregular field distribution and partial discharge activity due to electrode irregularities could be curtailed by the implementation of a solid dielectric layer on the electrode. However, the implementation of a solid dielectric layer in an FGIL would result in its reduced structural flexibility, which is objectionable regarding their metropolitan applications. A probable solution for conductor irregularity suppression in FGILs could be the implementation of a flexible semiconducting film (SCF) over the stranded conductor. SCFs basically exhibit non-linear conducting characteristics and will facilitate the minimization of the field irregularity and field intensity of FGILs without compromising their structural flexibility. Thus, considering the field irregularity concerns of FGILs, in this research, Autodesk Inventor® was used to model the geometric variants of stranded conductors. These conductor models were then analyzed in COMSOL Multiphysics® regarding electrostatic and dielectric aspects along with the development of the geometrically and electrostatically optimized FGIL model. Considering the performed optimized design, an 11 kV scaled-down model of a 132 kV FGIL was also fabricated in order to practically investigate the electrostatic and dielectric stresses in the FGIL through a high-voltage experimental setup. Simulation and experimental investigations revealed that SCF-coated stranded conductor significantly minimized the field irregularity of the FGIL and improved its dielectric breakdown characteristics.

2. Stranded Conductor Geometric Variants

Stranded conductors are normally discriminated on the basis of strand geometry as well as the compactness technique used in the conductor development [57]. The geometric configuration of stranded conductors used in UGCs and OHLs are specified in Table 1, and Figure 1 represents circular and trapezoidal strand conductors [57].

Table 1. Stranded conductors used in conventional power transmission systems.

Sr. No.	Conductor Type	Strand Geometry
1.	Concentric strand	Circular
2.	Compact strand	Circular
3.	Compact strand	Trapezoidal

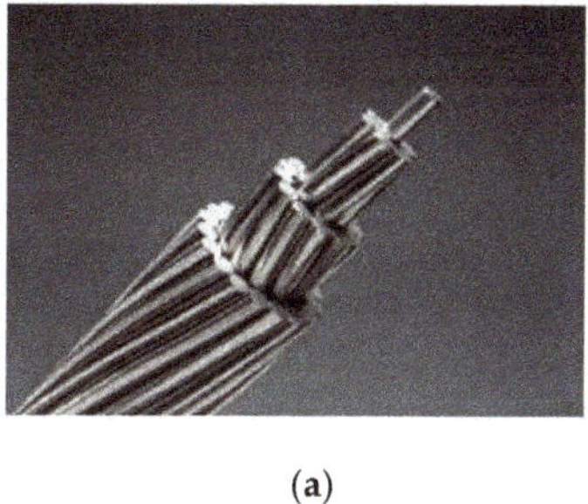

(a)

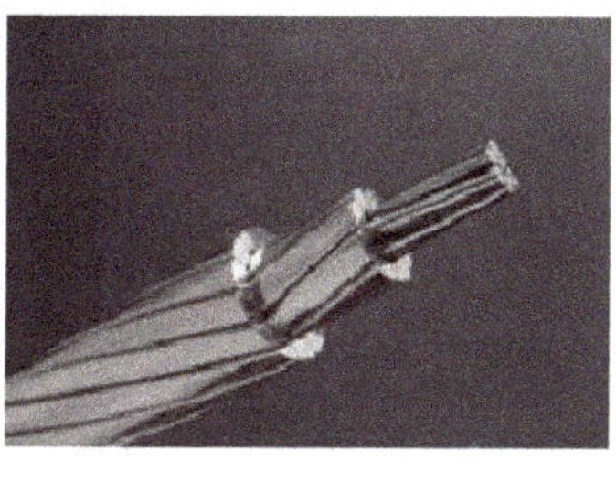

(b)

Figure 1. (a) Circular strand and (b) trapezoidal strand conductors used in conventional power transmission schemes.

Electric Field Dispersal Regarding Strand Geometry

The electric field dispersion in a region normally depends upon electrode geometry, surface irregularity, and gap distribution [58]. The field utilization factor (FUF) gives an idea of the effective utilization of field space and facilitates analysis of the electrostatic stresses imposed upon the dielectric material. In general, the FUF can be evaluated through Equation (1), where E_{avg} denotes the average electric field and E_{max} denotes the maximum electric field. Further, pertaining to its coaxial configuration, the FUF for an FGIL can be calculated through Equation (2), where R is the enclosure's radius in millimeters and r is the conductor's radius in millimeters [41].

$$\eta = \frac{E_{avg}}{E_{max}} \tag{1}$$

$$F = \frac{r \cdot \left(\ln \frac{R}{r} \right)}{R - r} \tag{2}$$

3. Design and Analysis

3.1. Dimensional Optimization of FGIL Enclosure Apropos of Stranded Conductor

The selection of a stranded conductor for an FGIL requires reconsideration regarding enclosure diameter because the FUF for GILs is normally kept in the range of 0.5 to 0.6 and is directly related with their dimensional specifications [24,41,42]. In a standard GIL, enclosure and conductor dimensions are normally selected to have approximately 1 as the solution of the logarithmic expression in Equation (2). That is, the enclosure diameter is approximately three times the conductor diameter [24,41,42]. However, in order to have an optimized enclosure size regarding the required FUF, Equation (2) was rearranged for enclosure dimension and expressed as Equation (3). Considering that Equation (3) appears as an implicit equation, its solution was performed through the Newton–Raphson iterative (NR) method in MATLAB® with the required accuracy up to four decimals and an initial estimate of 50 for the unknown parameter (i.e., enclosure radius). The estimated values and their errors showed a converging trend, and the enclosure radius finally converged in eleven iterations up to the required accuracy.

Dimensional appraisal of conductor and enclosure revealed that the enclosure was approximately three times the conductor size and resulted in the achievement of the desired FUF.

$$R = rexp\left(F\cdot\left(\left(\tfrac{R}{r}\right)-1\right)\right) \tag{3}$$

3.2. Electrostatic Field Optimization of the FGIL

Electric field optimization is obligatory in electrical systems in order to eradicate the prospect of dielectric failure due to partial discharge or gap discharge [58]. Considering the stranded conductor as a potential candidate for pliable GILs and concerning its surface irregularity, detailed electrostatic appraisal is essentially required for the proposed scheme. Protrusions and surface irregularities in stranded conductors may lead to escalated electric fields on the conductor's surface contour, which may result in detrimental partial discharge activity followed by dielectric strength degradation due to streamers [43–45,58]. Thus, considering the importance of field dispersion in pliable GIL, COMSOL Multiphysics®-based electrostatic analysis was performed for the FGIL regarding the stranded conductor specimens given in Table 2 in comparison to existing GILs in order to achieve minimal electrostatic stresses as per the required standards for gas-insulated equipment. The stranded conductors used in the electrostatic examination were developed using Autodesk Inventor®. Dimensional and technical specifications like electrode gap, thickness, and diameter for the conventional and proposed schemes were based upon ASTM B 232, ASTM B 857, and 132 kV GIL standards along with the evaluations of Section 3.1 [59,60]. Table 2 presents the detailed specifications of different conductor specimens used in the electrostatic stress investigation [59,61,62].

Table 2. Conductor specimens used in the comparative appraisal.

Specimen No.	Category	Material	Structure	Strand Geometry	Profile	Diameter (mm)
1.	Conventional	Aluminum	Hollow		Smooth	89
2.	Proposed	Aluminum	Stranded	Circular	Irregular	44.79
3.	Proposed	Aluminum	Stranded	Trapezoidal	Irregular	44.70

3.2.1. Electrostatic Field Dispersion Apropos of Conventional and Stranded Conductors

Concerning the analysis of the field dispersion along with identification of regions of high electric fields in the proposed GIL scheme, COMSOL Multiphysics®-based models for conventional and pliable GILs were developed and compared regarding the different conductor configurations given in Table 2. Figure 2a,b demonstrates the electric potential and electric field dispersion in a conventional GIL. Figure 3a,b exhibits the electric potential and electrostatic field dispersal in the proposed GIL with a circular strand conductor. Figure 4a,b represents the electric potential and electrostatic field distribution in the proposed GIL with trapezoidal strand conductor. Field dispersion regarding conventional and proposed schemes revealed that stranded conductors resulted in regions of concentrated electric field on the conductor's surface contour. Figure 5a,b represents the enlarged view of such concentrated electric field regions in the FGIL scheme regarding specimen 2 and specimen 3 of Table 2. Critical perusal of Figures 2–5 regarding electric field dispersion reveals that due to protrusions and surface irregularities of the stranded conductors, high electric fields appeared on their surface contour as compared to the conventional scheme with a smooth solid conductor. However, the trapezoidal strand conductor had approximately 10% lower magnitude of maximum electric field stresses due to its relatively smoother profile in comparison to the circular strand conductor. Figure 6 compares the average and maximum electric fields for conventional and proposed GIL schemes regarding the different conductor specimens described in Table 2. Further, Figure 7 compares the FUF for conventional and proposed GIL schemes regarding the different conductor specimens described in Table 2. Detailed analysis of Figures 6 and 7 revealed that the surface irregularity of stranded conductors in the proposed pliable GIL resulted in objectionably high electric fields regarding specimen 2 and specimen 3 in comparison to the conventional scheme regarding specimen 1. Further, the field utilization factor was also reduced by 31% and 23% respectively for specimens 2 and 3 regarding proposed pliable GIL in comparison to specimen 1 regarding the conventional scheme. A probable solution to the above stated problem could be to enlarge the enclosure's diameter or to suppress the conductor's irregularity [63–65]. COMSOL Multiphysics®-based simulations were performed for this purpose, which revealed that enclosure enlargement resulted in the minimization of the irregular field distribution and reduced the electrostatic stresses on the conductor's surface. However, the FUF was reduced in comparison to the standard allowable limit for GILs because as per GIL standards, the enclosure's diameter should be approximately three times the conductor's diameter in order to acquire an FUF in the permissible range of 0.5 to 0.6 [17,24,41,42]. The violation of the aforementioned constraint regarding enclosure diameter resulted in a poor field utilization factor for the proposed scheme, which is objectionable as per GIL standards. Thus, remedial measures regarding suppression of irregularities in the stranded conductor must be taken in order to achieve the required FUF and eradicate concentrated electric field regions.

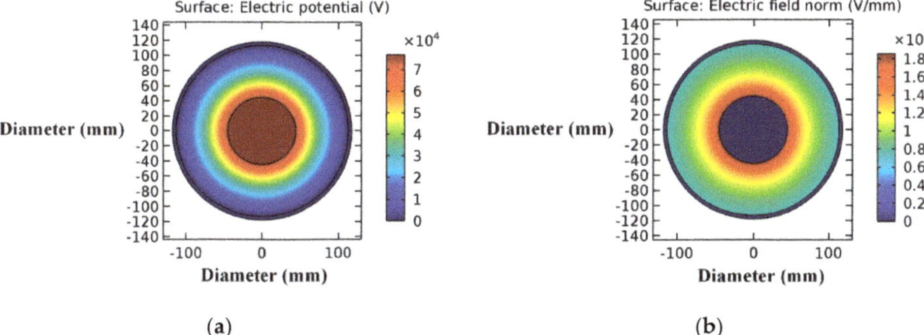

(a)   (b)

Figure 2. (a) Potential difference and (b) field distribution apropos of a conventional gas-insulated transmission line (GIL) regarding specimen 1 of Table 2.

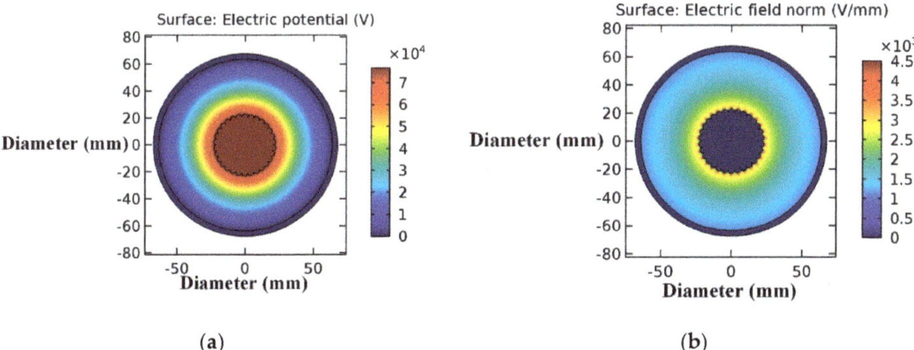

(a)   (b)

Figure 3. (a) Potential difference and (b) field distribution apropos of the proposed pliable GIL regarding specimen 2 of Table 2.

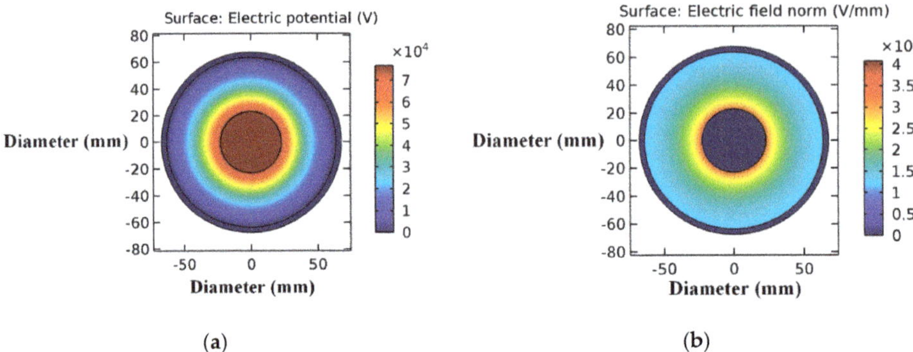

(a)

(b)

Figure 4. (**a**) Potential difference and (**b**) field distribution apropos of the proposed pliable GIL regarding specimen 3 of Table 2.

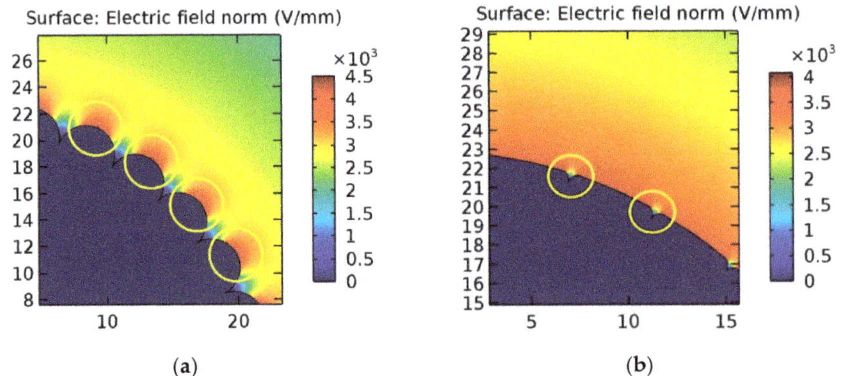

(a)

(b)

Figure 5. Location and magnitude of the maximum electric field for the proposed pliable GIL regarding (**a**) specimen 2 and (**b**) specimen 3 of Table 2.

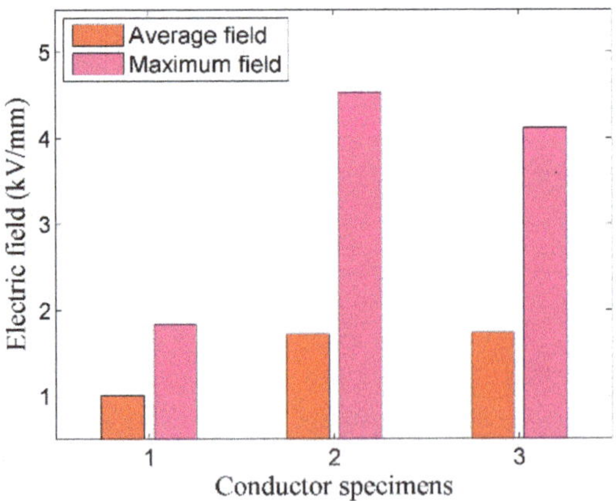

Figure 6. Average and maximum electric field comparison regarding the different conductor specimens described in Table 2.

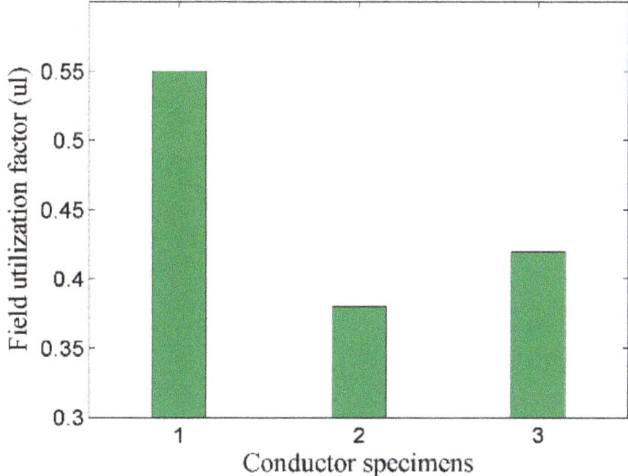

Figure 7. Field utilization factor comparison regarding the different conductor specimens described in Table 2.

3.2.2. Contour Irregularity Suppression of Stranded Conductor

Considering the objectionable deviations in the field utilization of the proposed FGIL due to stranded conductors, irregularity suppression essentially needs to be done in order to acquire the required FUF. A probable solution could be the implementation of a silicon carbide (SiC)-impregnated polyester-based SCF of 0.1–0.4 mm thickness on the stranded conductor in order to acquire a relatively smoother conductor profile [39,66,67]. The implementation of such film-coated stranded conductors in gas-insulated equipment necessitates detailed electrostatic and dielectric appraisal, as no published research regarding the implementation of field-graded stranded conductors in gas-insulated equipment exists to date.

3.2.3. Electrostatic Field Dispersion Apropos of Film-Coated Stranded Conductors

Concerning the effectivity of irregularity suppression for stranded conductors in terms of field utilization factor and electric field dispersion, SCF-coated stranded conductors were developed using Autodesk Inventor®. Dimensional specifications for the SCF-coated stranded conductors were based upon the ASTM B 232 and ASTM B 857 standards for stranded conductors, and the film thickness was based upon the standard film thickness for power cables [57,61,68]. Detailed specifications of the developed film-coated stranded conductors along with conventional GIL conductor are given in Table 3. Considering the conductor specimens given in Table 3, COMSOL Multiphysics®-based pliable GIL models were developed and analyzed in comparison to existing GIL schemes so as to achieve the desired electrostatic performance per the standards for GILs. Figure 8a,b demonstrates the electric potential and electrostatic field dispersion in the proposed pliable GIL scheme regarding specimen 2 of Table 3 respectively. Figure 9a,b exhibits the electric potential and electrostatic field distribution in the proposed pliable GIL scheme regarding specimen 3 of Table 3 respectively. Figure 10a,b shows the enlarged view of high electric field regions in the proposed FGIL scheme regarding specimens 2 and 3 of Table 3 respectively. Critical perusal of Figures 5 and 10 reveals that surface irregularity suppression resulted in substantial reduction in electrostatic stresses on the surface contour of the stranded conductor, and improved the field distribution for both stranded specimens of Table 3. However, specimen 3 had approximately 6% lower magnitude of maximum electrostatic stresses due to its nearly circular profile in comparison to specimen 2. Figure 11 compares the average and maximum electric fields for the conventional and proposed pliable GIL schemes regarding the respective conductor specimens of Table 3. Further, Figure 12 compares the field utilization factor of conventional and proposed GIL schemes regarding the respective conductor specimens of Table 3. Detailed analysis of

Figure 12 reveals that surface irregularity suppression resulted in achieving a relatively better FUF, with a trivial deviation of 7.5% and 1.8% regarding specimens 2 and 3 of Table 3 as compared to specimen 1 of the respective Table. In addition, a smoother conductor profile due to the SCF coating also resulted in substantial electrostatic stress reductions in the conductor contour up to 23% and 21% regarding specimens 2 and 3 of Table 3 in comparison to respective simple stranded conductors of Table 2. Further, in comparison to aluminum conductor steel reinforced (ACSR) and all aluminum alloy conductor (AAAC), due to their compact design, trapezoidal stranded conductors of the aluminum conductor steel supported (ACSS) category exhibit higher ampacity and thermal ratings within the same dimensional specifications [62]. Thus specimen 3 of Table 3 could serve as the optimal candidate regarding thermal, ampacity, and electrostatic requirements along with the desired flexibility for the proposed pliable GIL.

Table 3. Conductor specimens used in the comparative appraisal.

Specimen No.	Category	Material	Structure	Strand Geometry	Diameter (mm)	Film Material	Film Thickness (mm)
1.	Conventional	Aluminum	Hollow		89		
2.	Proposed	Aluminum	Stranded	Circular	44.79	SiC-impregnated polyester tape	0.2
3.	Proposed	Aluminum	Stranded	Trapezoidal	44.70	SiC-impregnated polyester tape	0.2

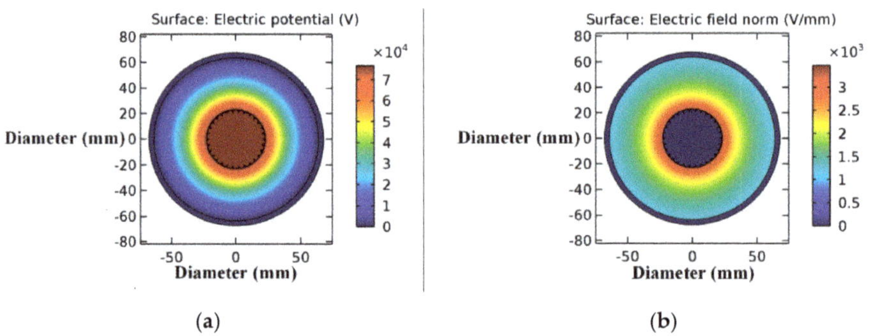

Figure 8. (**a**) Potential difference and (**b**) field distribution apropos of the proposed pliable GIL regarding specimen 2 of Table 3.

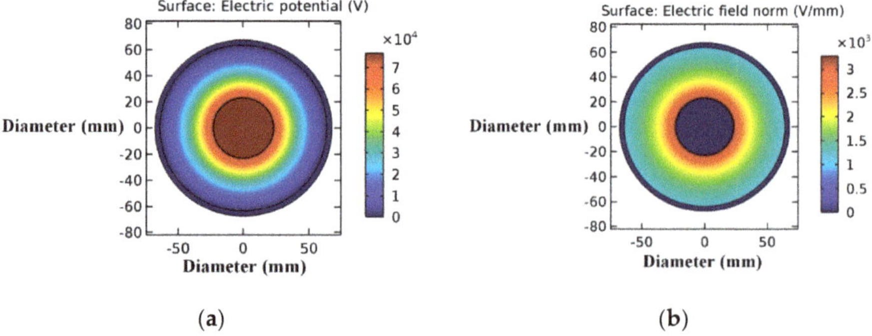

Figure 9. (**a**) Potential difference and (**b**) field distribution apropos of the proposed pliable GIL regarding specimen 3 of Table 3.

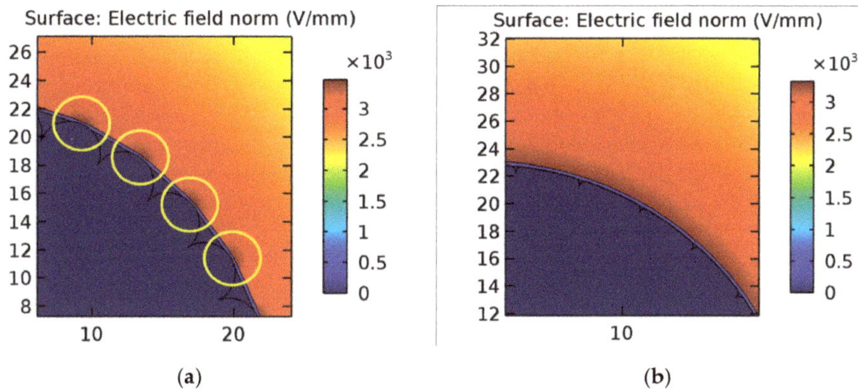

Figure 10. Location and magnitude of the maximum electric field for the proposed pliable GIL regarding (**a**) specimen 2 and (**b**) specimen 3 of Table 3.

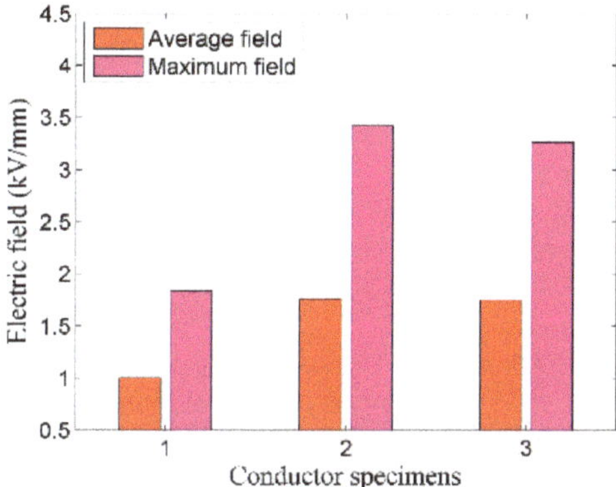

Figure 11. Average and maximum electric field comparison regarding the different conductor specimens described in Table 3.

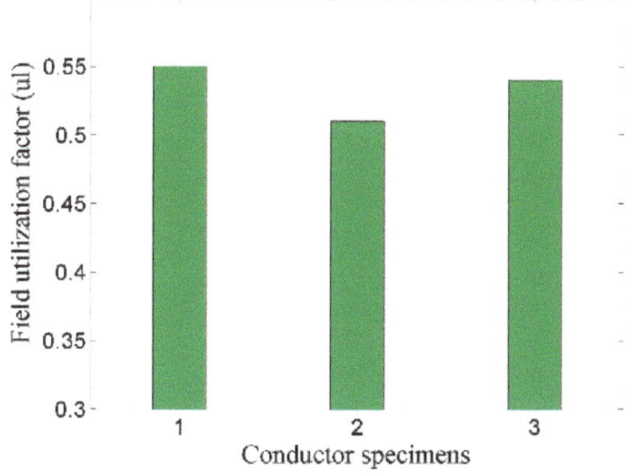

Figure 12. Field utilization factor comparison regarding the different conductor specimens described in Table 3.

3.3. Dielectric Appraisal of FGIL Apropos of the Stranded Conductor

In order to analyze the breakdown characteristics of the dielectric medium in the stranded-conductor-based FGIL model developed above, an analysis was performed regarding its minimum discharge voltage (i.e., breakdown voltage, BV). Existing methodologies for discharge voltage calculation can be categorized on the basis of the streamer breakdown theory as well as the critical field strength evaluations [58,69]. Both methodologies were incorporated in this research in order to ascertain the practicability of the suggested conductor scheme for a pliable GIL.

3.3.1. Breakdown Voltage of the Proposed Configuration Regarding Streamer Breakdown Theory

Per the streamer breakdown theory, a streamer may result in a partial discharge such as a corona or a complete gap discharge, and the associated potential level is considered as the breakdown voltage [58,69]. The minimum breakdown voltage in SF_6 insulated equipment can be evaluated by using Equation (4), where BV is the breakdown voltage in kV, P is the gas pressure in kPa, and d is the electrode gap in centimeters [58,69]. However, the effect of electrode surface irregularity should also be considered, as it significantly degrades the minimum breakdown voltage. Considering the conductor's surface irregularity in the model developed above, Equation (5) can be used for the evaluation of the minimum breakdown voltage, where BV is the breakdown voltage in kV, C is the curvature factor, d is the electrode gap in centimeters, F is the field utilization factor, P is the gas pressure in kPa, and S is the electrode roughness factor [58]. Further, considering the case when the dielectric medium comprises sulfur hexafluoride and nitrogen gases at a ratio of 20:80, its breakdown strength will be lesser in comparison to pure SF_6 gas, and will depend upon the percentage of SF_6 in the mixture. Thus, Equation (6) can be used for the evaluation of possible degradation in the minimum breakdown voltage of the gas mixture [69]. Figure 13 shows the electrode gap comparison between the 132-kV conventional and proposed GIL scheme which is further used in the evaluation of the minimum breakdown voltage for the proposed scheme. Reduction in electrode gap resulted due to the reduction of conductor diameter from 89 mm to 44.5 mm, as the proposed scheme comprises a stranded aluminum conductor whereas the conventional scheme comprises a hollow aluminum conductor. However, the ampacity of both conductors was kept approximately the same. Further, the diameter of the ground electrode was also reduced from 226 to 127.2 mm, as it is based on the dimensional evaluations performed in Section 3.1 regarding the standard field utilization factor as well as the standard dimensional specifications for GILs. Figure 14 represents the breakdown voltage for 100% SF_6 content and the respective reduction in this breakdown voltage due to surface irregularity of the stranded conductor using Equations (4) and (5). Moreover, it also highlights the reduction in breakdown voltage due to the reduced SF_6 content in the SF_6 and N_2 gas mixture through Equation (6). Critical analysis of Figures 13 and 14 shows that the BV for the given dimensional and operational specifications of the proposed 132 kV pliable GIL was well above the normal operating voltage and the standard basis insulation level (BIL) value of 132 kV. Thus, per the evaluated dimensional specifications, the proposed FGIL scheme exhibits good dielectric withstand capability.

$$BV = 1.321 \cdot (Pd)^{0.915} \tag{4}$$

$$BV = 0.8775 \cdot F \cdot S \cdot C \cdot P \cdot d \tag{5}$$

$$m = 38.03 \cdot n^{0.21} \tag{6}$$

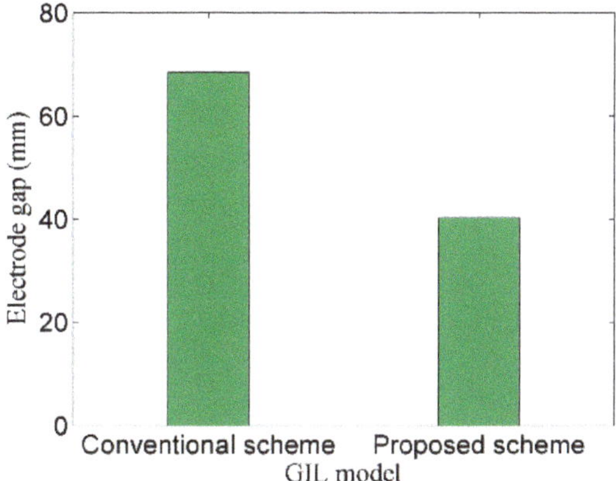

Figure 13. Electrode gap comparison between conventional and proposed flexible GIL (FGIL) schemes.

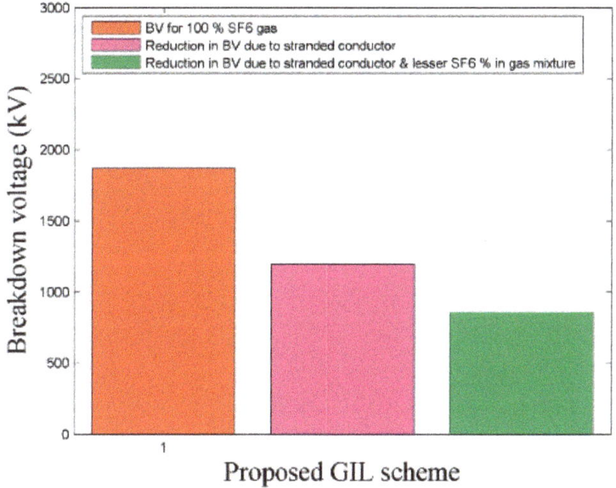

Figure 14. Breakdown voltage (BV) appraisal for the proposed FGIL scheme.

3.3.2. Breakdown Field Strength of the Proposed Configuration Regarding Critical Field Intensity Theory

In the critical field intensity method, the breakdown voltage of a dielectric gas is associated with critical field, electrode gap, FUF, electrode surface irregularity, and gas pressure [69]. According to this theory, the operational and design magnitudes of the electric field should be well below its critical value in order to avoid dielectric breakdown through avalanche [58]. Further, in accordance with GIL standards, the typical allowable design criterion regarding electric field strength is approximately 20 kV/mm, and might be higher such that the influenced region is not substantially enormous [24]. Thus, concerning the practical viability of the proposed 132-kV FGIL scheme, its electrostatic field appraisal regarding operational, design, and critical field values is essentially required. The operational and design values of the electric field in the proposed pliable GIL could be evaluated through Equation (7) by considering the normal operating voltage and standard BIL voltages, respectively [41]. Further, Equation (7) can be rearranged by considering Equations (1), (5), and (6) for the evaluation of the critical electric field as shown in Equation (8) [41,58]. In Equation (8), BV is the breakdown voltage and E_C is the critical electric field as evaluated on the basis of Equations (5) and (6). Figure 15 shows the comparison regarding the operational, design, and critical electric fields for the proposed scheme. Critical perusal of Figure 15 shows that the operational and design values for the electric field were well within limits as specified by the standards for gas-insulated equipment, and both field magnitudes were much less than the critical field value. Thus, per the appraised dimensional specifications, the proposed 132-kV FGIL scheme exhibits good electrostatic stress withstand capability.

$$E_{max} = \frac{U_0}{r \cdot \ln\left(\frac{R}{r}\right)} \tag{7}$$

$$E_c = \frac{BV}{f \cdot (R-r)} \tag{8}$$

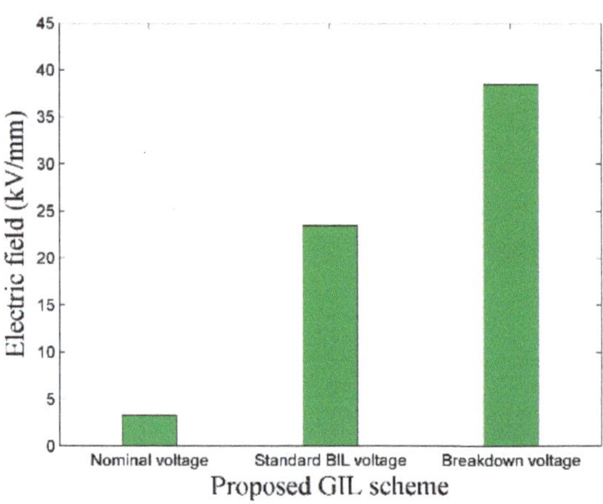

Figure 15. Electric field appraisal for the proposed FGIL scheme regarding operational, standard BIL, and breakdown voltages.

3.4. Field Stress Distribution in Bended Segment of FGIL

In order to analyze the practicability of the proposed scheme for bent segments, simulation regarding electrostatic stress distribution for FGIL in case of bend was performed in comparison to its equivalent straight model. For the said purpose, 20-m-long straight and bent FGIL models were developed using Autodesk Inventor®, and further field distribution and FUF for the two models were analyzed using COMSOL Multiphysics®. Line bending was performed as per the permissible longitudinal minimum bending radius (LMBR) of reinforced polyvinyl chloride (RPVC), and field distribution as well as FUF analysis for the respective FGIL models was performed regarding axial and radial cross sections. Critical appraisal of bent and straight FGIL models regrading axial cross section revealed a trivial deviation of 0.7% and 0.8% in electric field intensity and FUF, respectively. Moreover, detailed comparison of two FGIL models regrading radial cross section showed a slight deviation of 0.4% and 0.5% in electric field intensity and FUF, respectively. Minimal deviation in the compared models was observed because longitudinal bending as per LMBR limits resulted in negligible circumferential deformation as well as gradual bending. Thus, field magnitude and stress distribution for the bent line segment were nearly the same as those for the straight line segment.

4. Fabrication of the Scaled FGIL Model

Regarding the practical viability of the proposed scheme, an 11-kV scaled-down model of the 132-kV FGIL was fabricated on the basis of electrostatic modeling by replicating the field distribution of a high-voltage GIL for a scaled down model [17,24,41,42]. Considering the standard FUF and allowable maximum electric field for the 132 kV GIL, dimensional specifications regarding enclosure and conductor of the 11 kV scaled down model were evaluated per the technique described in Section 3.1. After finalizing the dimensional specifications, the scaled down model was first analyzed and compared with the actual 132 kV GIL model by using the technique described in Sections 3.2 and 3.3 regarding field distribution, field magnitude, FUF, and breakdown characteristics using COMSOL Multiphysics®. Then, a practical model was developed in order to conduct experimental investigations. RPVC was used as the enclosure material, and braided metallic mesh covered with aluminum foil was placed inside the enclosure as the ground terminal. A stranded aluminum conductor was placed inside the RPVC pipe, and threaded Teflon corks were used to prevent any gas leakage from pipe ends. Further, metallic clamps were placed on Teflon corks in order to avoid cork slippage and gas leakage at high gas pressures. Electrically pretested open cell rebond foam of 105 kg/m^3 density was used to achieve concentric conductor alignment inside the enclosure [34]. A gas charging and discharging system was implemented to pressurize the flexible GIL model at different gas pressures, and to create vacuum. The material and thickness of SCF were selected per the scheme described in Sections 3.2.2 and 3.2.3. The specified FGIL model for high-voltage experimentation was developed for simple as well as SCF-wrapped stranded conductors. Figure 16a,b shows the simple and SCF-coated stranded conductors used in the development of the FGIL models respectively. Figure 17a shows the dimensional specifications of the designed flexible GIL model, and Figure 17b shows the fully developed model.

(a) (b)

Figure 16. (**a**) Simple stranded conductor and (**b**) semiconducting film (SCF)-coated stranded conductor used in the development of FGIL models.

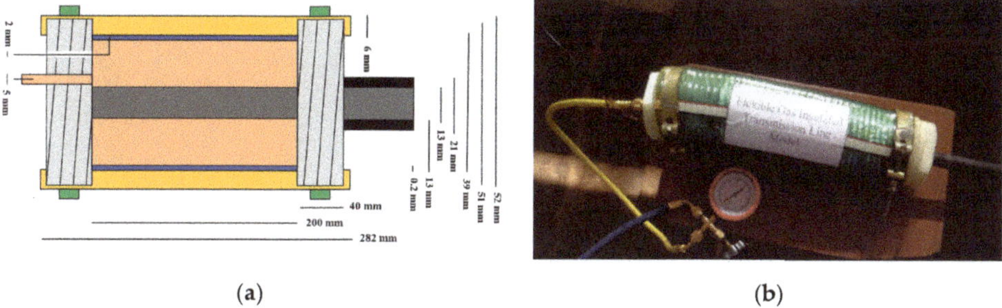

(a) (b)

Figure 17. (**a**) Dimensional specifications of the scaled FGIL model and (**b**) the fully developed scaled FGIL model.

5. Experimental Setup Development

Pertaining to the practicability of the proposed scheme, an experimental setup was developed regarding lightning impulse and power frequency disruptive discharge tests for the proposed scheme, as per the IEC 60060-1:2010 standard [70]. FGIL models, fabricated respectively with simple and SCF-coated stranded conductors, were used in this experimental investigation. Concerning the lightning impulse discharge tests, U_{50} for different GIL specimens was determined by Up–Down method, where for power frequency discharge an average of ten disruptive discharges was considered. A compressor and a pressure control unit were incorporated in order to create a vacuum in the developed GIL model along with the injection of dielectric gas at the desired pressure. Gas-insulated equipment normally utilizes pure SF_6 or a mixture of SF_6 and N_2 at a ratio of 20:80, but the required gas pressure in the latter case was almost doubled as compared to the prior case. Considering the security concerns associated with the high-pressure containment of SF_6/N_2 mixture at the laboratory level, pure SF_6 gas was used, as it would result in a significant reduction of the required gas pressure without compromising the insulation characteristics. A block diagram of the experimental setup is shown in Figure 18. Figure 19a represents the different components used in the high-voltage experimentation, while Figure 19b represents the experimental setup placed in a high-voltage laboratory.

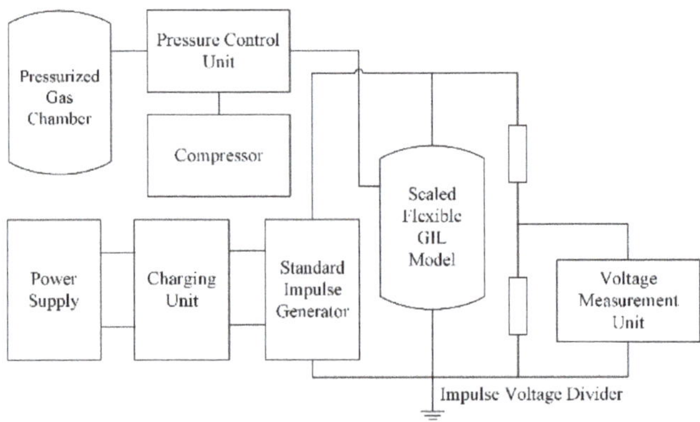

Figure 18. Experimental setup for the dielectric strength testing of the developed FGIL models.

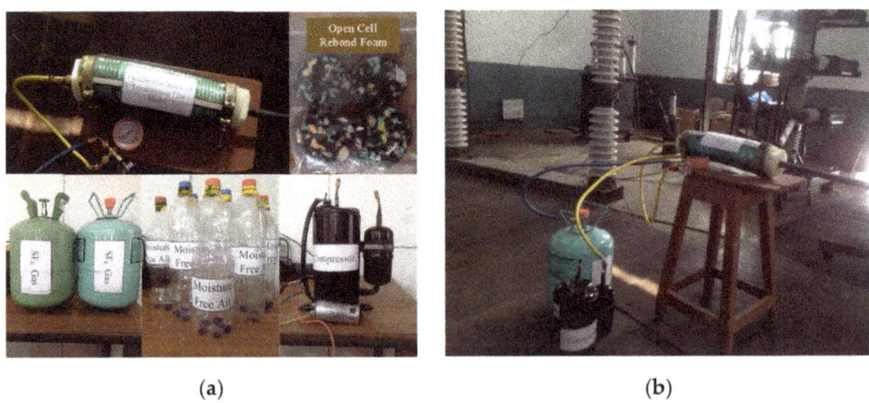

(a)   (b)

Figure 19. (**a**) Different components used in the discharge tests and (**b**) the experimental setup for lightning impulse and disruptive discharge tests.

5.1. Dielectric Breakdown Analysis of the Fabricated FGIL Models

Experimental analysis apropos of the dielectric breakdown of the developed FGIL models was performed for SF_6 and air gases at different gas pressures in order to investigate the power frequency and impulse discharge characteristics of the FGILs.

5.1.1. Power Frequency Discharge Test

In order to appraise the dielectric characteristics of the FGILs regarding simple and film-coated stranded conductors, power frequency discharge tests using air and SF_6 were performed for the fabricated FGIL models. After creating vacuum in the FGIL models, moisture-free air was filled at different gas pressures from 1 to 2.5 bar, and the discharge voltage was noted for both GIL models. Followed by air, similar power frequency discharge tests were performed regarding SF_6 gas for both FGIL specimens at different gas pressures from 1 to 2.5 bar. Figure 20 shows the disruptive discharge test results of air- and SF_6-filled simple and film-coated stranded-conductor-based FGIL models under different gas pressures. Critical analysis of Figure 20 reveals that the discharge voltage increased with increasing gas pressure in all cases. However, the FGIL model with a film-coated stranded conductor had relatively higher discharge voltages in air as well as sulfur hexafluoride respectively in comparison to the FGIL model with a simple stranded conductor. Further, pertaining to their higher dielectric strength, SF_6-filled FGIL models achieved higher breakdown voltages at the respective gas pressures.

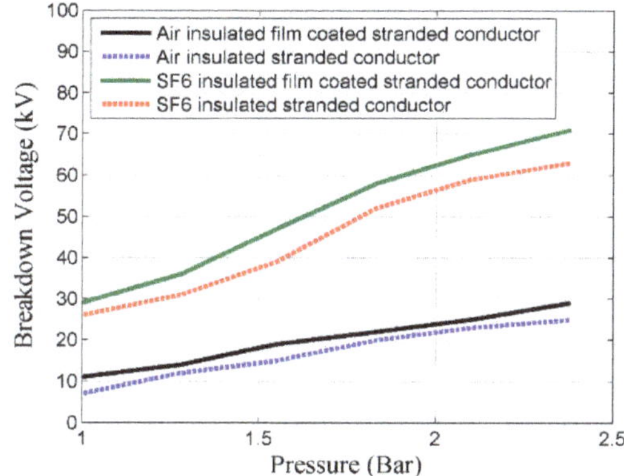

Figure 20. Power frequency disruptive discharge investigations regarding air and SF_6 for different FGIL models.

5.1.2. Lightning Impulse Discharge Test

Concerning the impulse withstand characteristics of the developed models, lightning impulse voltage tests were conducted for SF_6- and air-filled FGIL models with simple and SCF-coated stranded conductors. After creating vacuum in the FGIL models, moisture-free air was injected at different gas pressures from 1 to 2.5 bar, and lightning impulse discharge voltage was noted for both GIL models. Followed by air, similar lightning impulse discharge tests were performed regarding SF_6 gas for both FGIL models at different gas pressures from 1 to 2.5 bar. Figure 21 represents the test results of SF_6- and air-filled simple and film-coated stranded-conductor-based GIL models under different gas pressures. Critical analysis of Figure 21 reveals that the impulse discharge voltage increased with increasing gas pressure in all cases. However, the GIL model with a film-coated stranded conductor had relatively higher discharge voltages regarding air as well as sulfur hexafluoride respectively in comparison to the GIL model with a simple stranded conductor. Further, owing to its higher dielectric strength, SF_6-filled GIL models achieved the required BIL value for 11 kV beyond 2 bar pressure. Figure 22 shows the recorded waveforms of lightning impulse discharge tests regarding air and SF_6 gases at a pressure of 2.38 bar.

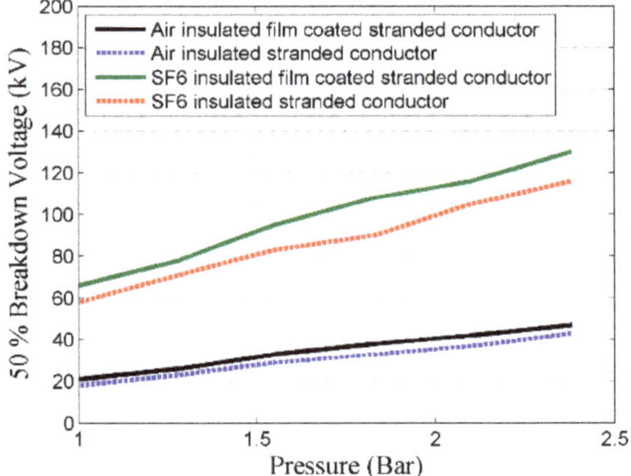

Figure 21. Lightning impulse discharge investigations regarding air and SF_6 for different FGIL models.

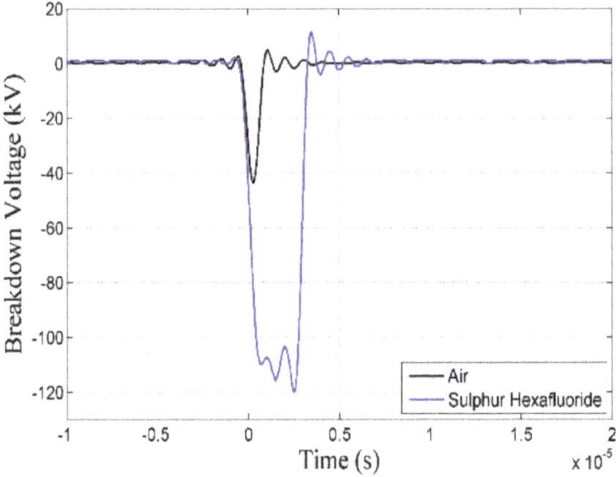

Figure 22. Measured lightning impulse discharge voltage regarding air and SF_6 in fabricated FGIL model at 2.38 bar.

5.2. Critical Field and Breakdown Field Analysis of Fabricated FGIL Models

Concerning the GIL, constraints regarding dielectric design require that the $(E/P)_{Breakdown}$ should be relatively lesser than the $(E/P)_{Critical}$ of the respective dielectric gas. The critical reduced field strength $(E/P)_{Critical}$ at $(\alpha - \eta) = 0$ regarding air and SF_6 per the computations using the BOLSIG+ tool were estimated as 30 kV/cm/bar and 89 kV/cm/bar [41]. Here, η represents the electron attachment rate and α is the coefficient of ionization. Regarding the developed experimental setup, the pressure normalized maximum field strength at U_{50}, $(E_{max}/P)_{Breakdown}$, could be evaluated by rewriting Equation (7) as Equation (9), where r represents the conductor's radius in mm, R represents the enclosure's radius in mm, and P is the gas pressure in kPa. Figure 23 shows the computations by Equation (9) regarding the experimental findings of lightning impulse discharge characteristics for air- and SF_6-insulated FGIL models with simple and SCF-coated stranded conductors. Critical analysis of Figure 23 reveals that the $(E/P)_{Breakdown}$ was lesser than the $(E/P)_{Critical}$ for all scenarios, and furthermore, the SCF coating over the stranded conductor enhanced the breakdown field level for both air- and SF_6-insulated FGIL models. Thus, the developed FGIL models fulfill the above-stated dielectric design requirements for GILs.

$$\left(E_{max}/P\right)_{Breakdown} = \frac{U_{50}}{r \cdot \ln\left(\frac{R}{r}\right) \cdot P}$$

(9)

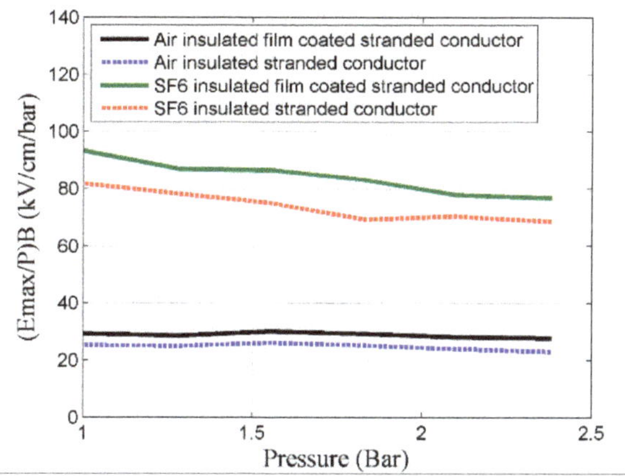

Figure 23. Field strength comparison in fabricated FGIL models regarding air and SF_6 at different gas pressures.

6. Conclusions

Conventional GILs are comprised of a hollow conductor which, owing to its intrinsic rigidity, restricts several application perspectives of conventional GILs specifically in metropolitan areas. Thus, the incorporation of structural flexibility in GILs is essential in order to curtail the operational intricacies of conventional GILs. In this research, FGIL models based on flexible simple stranded and flexible field graded stranded conductors were developed and analyzed regarding electrostatic and dielectric aspects through simulation and experimental assay.

Simulation results revealed that the simple stranded conductors had regions of objectionably high electric fields which ultimately resulted in 31% and 23% degradation of the FUF regarding circular strand and trapezoidal strand conductors respectively in comparison to the conventional GIL. Thus, simple stranded conductors may result in dielectric breakdown due to their surface irregularity, and require contour stress minimization. Possible solutions regarding stress minimization include enclosure enlargement and the suppression of conductors' irregularity. However, enclosure enlargement significantly deviated the FUF of the FGIL from its allowable range, which is highly objectionable according to GIL standards. Thus, field-graded stranded-conductor-based FGIL models were developed and analyzed through simulation and experimental investigations.

Simulation results revealed that SiC-coated stranded conductors resulted in the achievement of analogous electrostatic characteristics compared to the conventional GIL, with a trivial deviation of 7.2% and 1.8% in the FUF for circular strand and trapezoidal strand conductors, respectively, which are quite acceptable per the allowable FUF range for GILs. Further, critical comparison regarding dielectric aspects revealed that the breakdown voltage for the proposed scheme was approximately 23% above the required standard BIL value for GILs. In addition, electric fields for the proposed scheme regarding standard BIL voltage were approximately 38% below the critical field value and were well within the standard allowable range for electric fields in GILs.

Additionally, experimental investigations of fabricated FGIL models revealed that in comparison to the simple stranded-conductor-based model, the field-graded stranded-conductor-based model exhibited approximately 10–20% and 5–15% higher discharge voltages in power frequency and lightning impulse discharge tests. Moreover, the $(E/P)_{Breakdown}$ for the fabricated pliable models were observed to be relatively lesser than the $(E/P)_{Critical}$ at $(\alpha - \eta) = 0$ for the respective dielectric gases.

Consequently, simulation and experimental analysis revealed that the proposed conductor scheme could facilitate the achievement of the required dielectric and electrostatic characteristics for FGILs as described by GIL standards. However, the next step of this research is to perform similar high-voltage investigations on a full-scale 132-kV FGIL demonstrator.

Author Contributions: Conceptualization, T.I. and A.A.Q.; Data curation, M.J.A. and H.S.K.; Formal analysis, M.J.A.; Investigation, M.J.A. and A.S.; Methodology, M.J.A.; Project administration, T.I. and A.A.Q.; Software, M.J.A. and H.S.K.; Writing—original draft, M.J.A.; Writing—review and editing, T.I., A.A.Q., and A.S.

Funding: This research received no external funding.

Conflicts of Interest: The authors declare no conflict of interest.

References

1. Mohammadi, F.; Zheng, C. Stability Analysis of Electric Power System. In Proceedings of the 4th National Conference on Technology in Electrical and Computer Engineering, Tehran, Iran, 19 February 2018.
2. Adedeji, K.; Ponnle, A.; Abe, B.; Jimoh, A. Effect of increasing energy demand on the corrosion rate of buried pipelines in the vicinity of high voltage overhead transmission lines. In Proceedings of the 2015 Intl Aegean Conference on Electrical Machines & Power Electronics (ACEMP), 2015 Intl Conference on Optimization of Electrical & Electronic Equipment (OPTIM) & 2015 Intl Symposium on Advanced Electromechanical Motion Systems (ELECTROMOTION), Side, Turkey, 2–4 September 2015; pp. 299–303.
3. Sardaro, R.; Bozzo, F.; Fucilli, V. High-voltage overhead transmission lines and farmland value: Evidences from the real estate market in Apulia, southern Italy. *Energy Policy* **2018**, *119*, 449–457. [CrossRef]
4. Wiedemann, P.M.; Boerner, F.; Claus, F. How far is how far enough? Safety perception and acceptance of extra-high-voltage power lines in Germany. *J. Risk Res.* **2018**, *21*, 463–479. [CrossRef]
5. Fan, F.; Wu, G.; Wang, M.; Cao, Q.; Yang, S. Robot Delay-Tolerant Sensor Network for Overhead Transmission Line Monitoring. *Appl. Sci.* **2018**, *8*, 847. [CrossRef]
6. Gautam, B.R.; Li, F.; Ru, G. Assessment of urban roof top solar photovoltaic potential to solve power shortage problem in Nepal. *Energy Build.* **2015**, *86*, 735–744. [CrossRef]
7. Hajeforosh, S.; Pooranian, Z.; Shabani, A.; Conti, M. Evaluating the high frequency behavior of the modified grounding scheme in wind farms. *Appl. Sci.* **2017**, *7*, 1323. [CrossRef]
8. Mohammadi, F.; Zheng, C.; Su, R. Fault Diagnosis in Smart Grid Based on Data-Driven Computational Methods. In Proceedings of the 5th International Conference on Applied Research in Electrical, Mechanical, and Mechatronics Engineering, Tehran, Iran, 28 February 2019.
9. Xu, K.; Zhang, X.; Chen, Z.; Wu, W.; Li, T. Risk assessment for wildfire occurrence in high-voltage power line corridors by using remote-sensing techniques: A case study in Hubei Province, China. *Int. J. Remote Sens.* **2016**, *37*, 4818–4837. [CrossRef]
10. Benato, R.; Sessa, S.D.; Guizzo, L.; Rebolini, M. Saving environmental impact of electrical energy transmission by employing existing/planned transport corridors. In Proceedings of the 2017 IEEE International Conference on Environment and Electrical Engineering and 2017 IEEE Industrial and Commercial Power Systems Europe (EEEIC/I&CPS Europe), Milan, Italy, 6–9 June 2017; pp. 1–6.
11. Benato, R.; Sessa, S.D.; Guizzo, L.; Rebolini, M. Synergy of the future: High voltage insulated power cables and railway-highway structures. *IET Gener. Transm. Distrib.* **2017**, *11*, 2712–2720. [CrossRef]
12. Lauria, D.; Quaia, S. Technical comparison between a gas-insulated line and a traditional three-bundled OHL for a 400 kV, 200 km connection. In Proceedings of the 2015 International Conference on Clean Electrical Power (ICCEP), Taormina, Italy, 16–18 June 2015; pp. 597–601.
13. Kukharchuk, I.; Kazakov, A.; Trufanova, N. Investigation of heating of 150 kV underground cable line for various conditions of laying. In Proceedings of the IOP Conference Series: Materials Science and Engineering, Tomsk, Russia, 4–6 December 2017; p. 022041.
14. Salata, F.; Nardecchia, F.; de Lieto Vollaro, A.; Gugliermetti, F. Underground electric cables a correct evaluation of the soil thermal resistance. *Appl. Therm. Eng.* **2015**, *78*, 268–277. [CrossRef]

15. Magier, T.; Tenzer, M.; Koch, H. Direct Current Gas-Insulated Transmission Lines. *IEEE Trans. Power Deliv.* **2018**, *33*, 440–446. [CrossRef]
16. Magier, T.; Dehler, A.; Koch, H. AC Compact High Power Gas-Insulated Transmission Lines. In Proceedings of the 2018 IEEE/PES Transmission and Distribution Conference and Exposition (T&D), Denver, CO, USA, 16–19 April 2018; pp. 1–5.
17. Koch, H.; Goll, F.; Magier, T.; Juhre, K. Technical aspects of gas insulated transmission lines and application of new insulating gases. *IEEE Trans. Dielectr. Electr. Insul.* **2018**, *25*, 1448–1453. [CrossRef]
18. Widger, P.; Haddad, A. Evaluation of SF6 leakage from gas insulated equipment on electricity networks in Great Britain. *Energies* **2018**, *11*, 2037. [CrossRef]
19. Brinkley, C.; Leach, A. Energy next door: A meta-analysis of energy infrastructure impact on housing value. *Energy Res. Soc. Sci.* **2019**, *50*, 51–65. [CrossRef]
20. Mueller, C.E.; Keil, S.I.; Bauer, C. Effects of spatial proximity to proposed high-voltage transmission lines: Evidence from a natural experiment in Lower Saxony. *Energy Policy* **2017**, *111*, 137–147. [CrossRef]
21. Crespi, C.M.; Vergara, X.P.; Hooper, C.; Oksuzyan, S.; Wu, S.; Cockburn, M.; Kheifets, L. Childhood leukaemia and distance from power lines in California: A population-based case-control study. *Br. J. Cancer* **2016**, *115*, 122. [CrossRef] [PubMed]
22. Shabangu, T.; Shrivastava, P.; Abe, B.; Adedeji, K.; Olubambi, P. Influence of AC interference on the cathodic protection potentials of pipelines: Towards a comprehensive picture. In Proceedings of the 2017 IEEE AFRICON, Cape Town, South Africa, 18–20 September 2017; pp. 597–602.
23. Zarchi, D.A.; Vahidi, B. Optimal placement of underground cables to maximise total ampacity considering cable lifetime. *IET Gener. Transm. Distrib.* **2016**, *10*, 263–269. [CrossRef]
24. Koch, H. *Gas. Insulated Transmission Lines (GIL)*; John Wiley & Sons: Hoboken, NJ, USA, 2011.
25. McDonald, J.D. *Electric Power Substations Engineering*; CRC press: Boca Raton, FL, USA, 2016.
26. Zhou, T. Study on Mechanical Properties of Transmission Pipe Materials. In Proceedings of the IOP Conference Series: Materials Science and Engineering, Sanya, China, 10–11 November 2018; p. 012090.
27. Tenzer, M.; Koch, H.; Imamovic, D. Underground transmission lines for high power AC and DC transmission. In Proceedings of the 2016 IEEE/PES Transmission and Distribution Conference and Exposition (T&D), Dallas, TX, USA, 3–5 May 2016; pp. 1–4.
28. Chakir, A.; Koch, H. Corrosion protection for gas-insulated transmission lines. In Proceedings of the IEEE Power Engineering Society Summer Meeting, Chicago, IL, USA, 21–25 July 2002; pp. 220–224.
29. Chakir, A.; Koch, H. Seismic calculations of directly buried gas-insulated transmission lines (GIL). In Proceedings of the IEEE/PES Transmission and Distribution Conference and Exhibition, Yokohama, Japan, 6–10 October 2002; pp. 1026–1029.
30. Imamovic, D.; Tenzer, M.; Koch, H.; Lutz, B. High power underground transmission lines. In Proceedings of the 9th International Conference on Insulated Power Cables, Paris, France, 21–25 June 2015.
31. Alvi, M.J.; Izhar, T.; Qaiser, A.A.; Afzaal, M.U.; Anjum, A.; Safdar, A. Pliability Assay of Conventional Gas Insulated Transmission Line and Flexible Gas Insulated Transmission Line Regarding Horizontal Directional Drilling Based Underground Cable Laying for Metropolitan Areas. In Proceedings of the 2018 IEEE International Conference on Environment and Electrical Engineering and 2018 IEEE Industrial and Commercial Power Systems Europe (EEEIC/I&CPS Europe), Palermo, Italy, 12–15 June 2018; pp. 1–5.
32. Liang, X.; Shen, Y.; Liu, Y.; Wang, J.; Gao, Y.; Li, S.; Wang, M.; Gao, S. Investigations on the basic electrical properties of Polyurethane foam material. In Proceedings of the 2015 IEEE 11th International Conference on the Properties and Applications of Dielectric Materials (ICPADM), Sydney, Australia, 19–22 July 2015; pp. 863–866.
33. Zhu, Y.; He, L.; Zhou, S.; Zhang, D.; Fang, J. Insulation performance of rigid polyurethane foam and its application in station post insulator. In Proceedings of the 2018 12th International Conference on the Properties and Applications of Dielectric Materials (ICPADM), Xi'an, China, 20–24 May 2018; pp. 970–973.
34. Alvi, M.; Izhar, T.; Qaiser, A. Proposed scheme of pliable gas insulated transmission line and its comparative appraisal regarding electrostatic and dielectric aspects. *Electronics* **2018**, *7*, 328. [CrossRef]
35. Guo, J.; Chen, H. Research on Horizontal Directional Drilling Locatingtechnology Based on Seismic Interference. In Proceedings of the 2019 Symposium on Piezoelectrcity, Acoustic Waves and Device Applications (SPAWDA), Harbin, China, 11–14 January 2019; pp. 1–4.

36. Wang, Y.; Wen, G.; Chen, H. Horizontal directional drilling-length detection technology while drilling based on bi-electro-magnetic sensing. *Sensors* **2017**, *17*, 967. [CrossRef]

37. Bascom, E.C.R.; Rezutko, J. Novel installation of a 138kv pipe-type cable system under water using horizontal directional drilling. In Proceedings of the 2014 IEEE PES T&D Conference and Exposition, Chicago, IL, USA, 14–17 April 2014; pp. 1–5.

38. Paun, A.; Frigura-Iliasa, F.M.; Frigura-Iliasa, M.; Andea, P.; Vatau, D. A FEW ASPECTS CONCERNING THE INFLUENCE OF THE ELECTRODES ON THE SF^ sub 6^ INSULATION DIELECTRIC BEHAVIOR. *Int. Multidiscip. Sci. Geoconf. Sgem Surv. Geol. Min. Ecol. Manag.* **2018**, *18*, 9–16.

39. Yang, X.; Zhao, X.; Hu, J.; He, J. Grading electric field in high voltage insulation using composite materials. *IEEE Electr. Insul. Mag.* **2018**, *34*, 15–25. [CrossRef]

40. Yang, X.; Zhao, X.; Li, Q.; Hu, J.; He, J. Nonlinear effective permittivity of field grading composite dielectrics. *J. Phys. D Appl. Phys.* **2018**, *51*, 075304. [CrossRef]

41. Chen, L.; Griffiths, H.; Haddad, A.; Kamarudin, M. Breakdown of cf 3 i gas and its mixtures under lightning impulse in coaxial-GIL geometry. *IEEE Trans. Dielectr. Electr. Insul.* **2016**, *23*, 1959–1967. [CrossRef]

42. Chen, L.; Widger, P.; Kamarudin, M.S.; Griffiths, H.; Haddad, A. CF3I gas mixtures: Breakdown characteristics and potential for electrical insulation. *IEEE Trans. Power Deliv.* **2016**, *32*, 1089–1097. [CrossRef]

43. Gao, C.; Yu, Y.; Wang, Z.; Wang, W.; Zheng, L.; Du, J. Study on the Relationship between Electrical Tree Development and Partial Discharge of XLPE Cables. *J. Nanomater.* **2019**, *2019*. [CrossRef]

44. Illias, H.A.; Tunio, M.A.; Bakar, A.H.A.; Mokhlis, H.; Chen, G. Partial discharge phenomena within an artificial void in cable insulation geometry: Experimental validation and simulation. *IEEE Trans. Dielectr. Electr. Insul.* **2016**, *23*, 451–459. [CrossRef]

45. Refaat, S.S.; Shams, M.A. A review of partial discharge detection, diagnosis techniques in high voltage power cables. In Proceedings of the 2018 IEEE 12th International Conference on Compatibility, Power Electronics and Power Engineering (CPE-POWERENG 2018), Qatar, Doha, 10–12 April 2018; pp. 1–5.

46. Danikas, M.; Papadopoulos, D.; Morsalin, S. Propagation of Electrical Trees under the Influence of Mechanical Stresses: A Short Review. *Eng. Technol. Appl. Sci. Res.* **2019**, *9*, 3750–3756.

47. Goetz, T.; Linde, T.; Simka, P.; Speck, J.; Backhaus, K.; Gabler, T.; Riechert, U.; Grossmann, S. Surface discharges on dielectric coated electrodes in gas-insulated systems under DC voltage stress. In Proceedings of the VDE High Voltage Technology 2018, ETG-Symposium, Berlin, Germany, 12–14 November 2018; pp. 1–6.

48. Romero, A.; Rácz, L.; Mátrai, A.; Bokor, T.; Cselkó, R. A review of sulfur-hexafluoride reduction by dielectric coatings and alternative gases. In Proceedings of the 2017 6th International Youth Conference on Energy (IYCE), Budapest, Hungary, 21–24 June 2017; pp. 1–5.

49. Wang, R.; Cui, C.; Zhang, C.; Ren, C.; Chen, G.; Shao, T. Deposition of SiO x film on electrode surface by DBD to improve the lift-off voltage of metal particles. *IEEE Trans. Dielectr. Electr. Insul.* **2018**, *25*, 1285–1292. [CrossRef]

50. Kumar, G.A.; Ram, S.T. Investigation of Dielectric Coating on Particle Movement in Single Phase Gas Insulated Bus Duct Under Sf 6/Co 2 Gas Mixture. In Proceedings of the 2018 International Conference on Emerging Trends and Innovations In Engineering And Technological Research (ICETIETR), Arakkunnam, Kerala, 11–13 July 2018; pp. 1–6.

51. Sun, Z.; Fan, W.; Liu, Z.; Bai, Y.; Geng, Y.; Wang, J. Improvement of dielectric performance of solid/gas composite insulation with YSZ/ZTA coatings. *Sci. Rep.* **2019**, *9*, 3888. [CrossRef] [PubMed]

52. Van Der Born, D.; Morshuis, P.; Smit, J.; Girodet, A. The influence of thin dielectric coatings on LI and AC breakdown strength in SF6 and dry air. In Proceedings of the 2013 IEEE International Conference on Solid Dielectrics (ICSD), Bologna, Italy, 30 June–4 July 2013; pp. 287–290.

53. Hama, H.; Okabe, S. Factors dominating dielectric performance of real-size gas insulated system and their measures by dielectric coatings in SF 6 and potential gases. *IEEE Trans. Dielectr. Electr. Insul.* **2013**, *20*, 1737–1748. [CrossRef]

54. Krivda, A.; Straumann, U.; Martinon, M.; Logakis, E.; Tsang, F.; Narendran, M. Electrical and chemical characterization of thin epoxy layers for high voltage applications. In Proceedings of the 2014 IEEE Conference on Electrical Insulation and Dielectric Phenomena (CEIDP), Des Moines, IA, USA, 19–22 October 2014; pp. 816–819.

55. Arnold, P.; Hering, M.; Tenbohlen, S.; Köhler, W.; Riechert, U.; Straumann, U. Discharge phenomena at partially damaged dielectric electrode coatings in SF6-insulated systems at AC and DC stress. In Proceedings of the 20th Internaional Conference on Gas Discharges and Their Application, Orléans, France, 6–11 July 2014.

56. Gremaud, R.; Doiron, C.; Baur, M.; Simka, P.; Teppati, V.; Hering, M.; Speck, J.; Grossmann, S.; Källstrand, B.; Johansson, K. Solid-gas insulation in HVDC gas-insulated system: Measurement modeling and experimental validation for reliable operation. In Proceedings of the Cigré Session 46, Paris, France, 21–26 August 2016; p. D1_101.

57. Conductor Strand Types. Available online: www.anixter.com (accessed on 1 April 2019).

58. Malik, N.H.; Al-Arainy, A.; Qureshi, M.I. *Electrical Insulation in Power Systems*; Marcel Dekker: New York, NY, USA, 1998.

59. Koch, H.J. Super Session 'Vision 2020'IEEE General Meeting 2008 Application of long high capacity gas insulated lines in structures. In Proceedings of the 2008 IEEE Power and Energy Society General Meeting-Conversion and Delivery of Electrical Energy in the 21st Century, Pittsburgh, PA, USA, 20–24 July 2018; pp. 1–5.

60. Benato, R.; Di Mario, C.; Koch, H. High-capability applications of long gas-insulated lines in structures. *IEEE Trans. Power Deliv.* **2007**, *22*, 619–626. [CrossRef]

61. Kopsidas, K.; Rowland, S.M.; Boumecid, B. A holistic method for conductor ampacity and sag computation on an OHL structure. *IEEE Trans. Power Deliv.* **2012**, *27*, 1047–1054. [CrossRef]

62. Alawar, A.; Bosze, E.J.; Nutt, S.R. A composite core conductor for low sag at high temperatures. *IEEE Trans. Power Deliv.* **2005**, *20*, 2193–2199. [CrossRef]

63. Christen, T.; Donzel, L.; Greuter, F. Nonlinear resistive electric field grading part 1: Theory and simulation. *IEEE Electr. Insul. Mag.* **2010**, *26*, 47–59. [CrossRef]

64. Donzel, L.; Greuter, F.; Christen, T. Nonlinear resistive electric field grading Part 2: Materials and applications. *IEEE Electr. Insul. Mag.* **2011**, *27*, 18–29. [CrossRef]

65. Gao, L.; Yang, X.; Hu, J.; He, J. ZnO microvaristors doped polymer composites with electrical field dependent nonlinear conductive and dielectric characteristics. *Mater. Lett.* **2016**, *171*, 1–4. [CrossRef]

66. Boucher, S.; Chambard, R.; Hassanzadeh, M.; Castellon, J.; Metz, R. Study of the influence of nonlinear parameters on the efficiency of a resistive field grading tube for cable termination. *IEEE Trans. Dielectr. Electr. Insul.* **2018**, *25*, 2413–2420. [CrossRef]

67. Pradhan, M.; Greijer, H.; Eriksson, G.; Unge, M. Functional behaviors of electric field grading composite materials. *IEEE Trans. Dielectr. Electr. Insul.* **2016**, *23*, 768–778. [CrossRef]

68. Power Cables. Available online: www.zmscable.com (accessed on 1 April 2019).

69. Li, X.; Zhao, H.; Jia, S.; Murphy, A.B. Study of the dielectric breakdown properties of hot SF6–CF4 mixtures at 0.01–1.6 MPa. *J. Appl. Phys.* **2013**, *114*, 053302. [CrossRef]

70. IEC Standard, I. *60060-1: High Voltage Test Techniques—Part 1: General Definitions and Test Requirements*; International Electrotechnical Commission: Geneva, Switzerland, September 2010.

Article

An Improved Mixed AC/DC Power Flow Algorithm in Hybrid AC/DC Grids with MT-HVDC Systems

Fazel Mohammadi [1],*, Gholam-Abbas Nazri [2] and Mehrdad Saif [1]

[1] Electrical and Computer Engineering (ECE) Department, University of Windsor, Windsor, ON N9B 1K3, Canada
[2] Electrical and Computer Engineering, Wayne State University, Detroit, MI 48202, USA
* Correspondence: fazel@uwindsor.ca or fazel.mohammadi@ieee.org

Received: 30 November 2019; Accepted: 19 December 2019; Published: 31 December 2019

Abstract: One of the major challenges on large-scale Multi-Terminal High Voltage Direct Current (MT-HVDC) systems is the steady-state interaction of the hybrid AC/DC grids to achieve an accurate Power Flow (PF) solution. In PF control of MT-HVDC systems, different operational constraints, such as the voltage range, voltage operating region, Total Transfer Capability (TTC), transmission reliability margin, converter station power rating, etc. should be considered. Moreover, due to the nonlinear behavior of MT-HVDC systems, any changes (contingencies and/or faults) in the operating conditions lead to a significant change in the stability margin of the entire or several areas of the hybrid AC/DC grids. As a result, the system should continue operating within the acceptable limits and deliver power to the non-faulted sections. In order to analyze the steady-state interaction of the large-scale MT-HVDC systems, an improved mixed AC/DC PF algorithm for hybrid AC/DC grids with MT-HVDC systems considering the operational constraints is developed in this paper. To demonstrate the performance of the mixed AC/DC PF algorithm, a five-bus AC grid with a three-bus MT-HVDC system and the modified IEEE 39-bus test system with two four-bus MT-HVDC systems (in two different areas) are simulated in MATLAB software and different cases are investigated. The obtained results show the accuracy, robustness, and effectiveness of the improved mixed AC/DC PF algorithm for operation and planning studies of the hybrid A/DC grids.

Keywords: improved mixed AC/DC power flow; multi-terminal high voltage direct current (MT-HVDC) systems; operational constraints; voltage-sourced converter (VSC)-high voltage direct current (HVDC) station

---

1. Introduction

Due to the recent developments in the power electronics technology, Voltage-Sourced Converter (VSC)-High Voltage Direct Current (HVDC) systems have solved the problem of bidirectional Power Flow (PF) in hybrid AC/DC grids [1–3]. MT-HVDC systems are capable of controlling the active and reactive power, independently. One of the important considerations to control the HVDC systems is that the V_{DC}-control and P-control VSC-HVDC stations should be capable of operating in inverter and rectifier modes, respectively [4–6]. The main purpose of applying different control strategies in MT-HVDC systems is to achieve a precise and secure control mode for MT-HVDC systems without violating the operational constraints. For stable operation and active and reactive PF, MT-HVDC systems need to maintain the DC voltage and frequency within the operating limits [1–4].

1.1. AC/DC Power Flow for Hybrid AC/DC Grids with MT-HVDC Systems

There have been some relevant surveys about power system operation considering PF problem solutions [7,8]. In traditional AC systems, the PF can be controlled through three hierarchical levels which are, HLI (generation level), HLII (generation and transmission levels), and HLIII (generation, transmission, and distribution levels). Considering hierarchical levels and the market analysis, the Independent Electricity System Operator (IESO) can only control the generated power on the power system, and based on that, control the PF to the next hierarchical level(s). The integration of MT-HVDC systems to the existing AC grids leads to increasing the region of controllability of the hybrid AC/DC grids. This fact is due to the capability of controlling active and reactive PF by each converter station in MT-HVDC systems [1,4,9]. Therefore, MT-HVDC systems can change the PF patterns, and from the IESO and Transmission System Operator (TSO) perspectives, these changes in the pattern may cause significant issues in hybrid AC/DC grids. Moreover, due to the temporary or permanent outages of the components in hybrid AC/DC grids, the dynamic behavior of nonlinear components can change the PF and enhance the risk of instability in the entire hybrid AC/DC grids [9–11].

1.2. Solving the Power Flow Problem for Hybrid AC/DC Grids with MT-HVDC Systems

The unified method and sequential method are the two well-known methods to solve the PF problem for hybrid AC/DC grids with MT-HVDC systems.

1.2.1. The Unified Method

The unified method solves the PF problem for the entire hybrid AC/DC grids using a modified Jacobian matrix [12–14], where all the AC and DC variables, such as the impact of DC links in the Jacobian matrix, can be calculated in each iteration process. Many techniques are proposed and developed to improve the efficiency of the unified method, but the main drawback of those methods is neglecting the impact of droop parameters' settings on the AC/DC PF [12–14]. Another issue of the unified method to solve the PF problem is that it needs an alternation of the extension of an existing AC PF.

1.2.2. The Sequential Method

The sequential method solves the AC/DC PF equations sequentially, one after another, in each iteration [15,16]. The main advantage of the sequential method is to make the solution easy to combine the DC PF to the AC PF solution, and it can be implemented easily when the extension of an existing AC PF is needed. In [17], a numerical method based on the Newton-Raphson algorithm to calculate the converter station losses is proposed. A detailed steady-state model of the converter station to solve the AC/DC PF problem sequentially based on the Newton-Raphson technique considering converter station losses and reference power is proposed in [18,19]. An algorithm with per-unit conversion and changing the bus numbers to simplify solving the PF problem when multiple DC lines and converter station outages occur is developed in [18]. The main drawback of this method is neglecting the AC grid connection in the problem. In [20], a detailed model of the converter station with AC/DC PF equations, including converter station losses, for solving the PF sequentially is proposed. In [21], a method to solve the PF problem through the Gauss-Seidel method is developed.

1.3. AC/DC Power Flow for MT-HVDC Systems Considering Droop Parameters

Controlling the droop parameters in MT-HVDC systems has a significant impact on the PF of AC and DC grids after an outage [4]. In [16], the concept of distributed DC voltage control with the droop parameters on the PF problem in MT-HVDC systems is considered. In [22], a methodology to determine the mean voltage instead of a single slack converter station in MT-HVDC systems to solve the DC PF is proposed. By interconnecting MT-HVDC systems to the large-scale AC grids, solving the AC/DC PF goes through a complicated process, and a combined solution of AC/DC PF considering all the system's variables and constraints is required.

To address all the above-mentioned challenges, a mixed AC/DC PF algorithm for hybrid AC/DC grids with MT-HVDC systems is proposed in this paper. The proposed strategy is a fast and accurate method, which is capable of optimizing the AC/DC PF calculations. Except for the high accuracy and optimized performance, considering all operational constraints and control objectives of the integration of MT-HVDC systems into the large-scale AC grids is the other contribution of this paper. The calculated results by the mixed AC/DC PF problem can be used for the planning, scheduling, state estimation, small-signal stability analyses. The mixed AC/DC PF algorithm is applied to a five-bus AC grid with a three-bus MT-HVDC system and the modified IEEE 39-bus test system with two four-bus MT-HVDC systems (in two different areas) which are simulated in MATLAB software. To check the performance of the mixed AC/DC PF algorithm, different cases are considered. The obtained results show the accuracy, robustness, and effectiveness of the mixed AC/DC PF algorithm.

2. Principles of Power Flow in Power Systems

2.1. AC Grid Power Flow

The main objective of the AC PF is to determine the magnitude and angle at each bus of the AC grids. In order to analyze the AC PF in power systems, the following assumptions are considered.

- The AC transmission networks have fast dynamics compared to the other components. In this regard, AC transmission networks can be represented by algebraic equations.
- Each transmission line and transformer is modeled by an equivalent π model.
- The power in AC grids is balanced.
- The positive sequence parameters on a per-phase basis are assumed.

Accordingly, the network equations can be written as follows:

$$
\begin{bmatrix}
\bar{I}_1 \\
\bar{I}_2 \\
\vdots \\
\bar{I}_i \\
\vdots \\
\bar{I}_{n_{AC}}
\end{bmatrix}
=
\begin{bmatrix}
\overline{Y}_{11} & \overline{Y}_{12} & \cdots & \overline{Y}_{1i} & \cdots & \overline{Y}_{1n_{AC}} \\
\overline{Y}_{21} & \overline{Y}_{22} & \cdots & \overline{Y}_{2i} & \cdots & \overline{Y}_{2n_{AC}} \\
\vdots & \vdots & \vdots & \vdots & \vdots & \vdots \\
\overline{Y}_{i1} & \overline{Y}_{i2} & \cdots & \overline{Y}_{ii} & \cdots & \overline{Y}_{in_{AC}} \\
\vdots & \vdots & \vdots & \vdots & \vdots & \vdots \\
\overline{Y}_{n_{AC}1} & \overline{Y}_{n_{AC}2} & \cdots & \overline{Y}_{n_{AC}i} & \cdots & \overline{Y}_{n_{AC}n_{AC}}
\end{bmatrix}
\begin{bmatrix}
\overline{V}_1 \\
\overline{V}_2 \\
\vdots \\
\overline{V}_i \\
\vdots \\
\overline{V}_{n_{AC}}
\end{bmatrix}
\tag{1}
$$

where $\bar{I}_i$ and $\overline{V}_i = V_i e^{j\theta_i}$ are the injected current and voltage at the ith node, $\overline{Y}_{ii}$ is the self-admittance at the ith node, $\overline{Y}_{ij}$ is the mutual admittance between nodes i and j, and n_{AC} represents the total number of buses in the AC grid.

The voltage of each node can be used for solving the equation of $[\bar{I}] - [\overline{Y_{bus}}][\overline{V}] = 0$. The injected current at the ith node is related to the injected power and bus voltage at that node and it can be calculated as follows:

$$
\bar{I}_i = \frac{P_i - jQ_i}{\overline{V}_i^*}
\tag{2}
$$

Considering the operational constraints in power systems, the PF problem becomes nonlinear, and the nodes in AC grids can be classified into four types:

- Slack bus where the voltage magnitude (V_i) and angle (θ_i) are determined.
- PV bus where the active power (P_i) injected to the grid and the voltage magnitude (V_i) are known.
- PQ bus where the active power and reactive power (Q_i) injected into the grid are known.
- MT-HVDC Point of Common Coupling (PCC) bus where different constraints based on the control mode of each converter station are applied.

From Equation (1), the injected current at the ith node (without considering the PCC buses of MT-HVDC systems) can be written as follows:

$$
\bar{I}_i = \sum_{j=1}^{n_{AC}} \overline{Y}_{ij}\overline{V}_j
\tag{3}
$$

Therefore, the active and reactive power injected to the grid can be derived as follows:

$$P_i = \sum_{j=1}^{n_{AC}} V_i V_j \left[G_{ij} \cos\left(\theta_i - \theta_j\right) + B_{ij} \sin\left(\theta_i - \theta_j\right) \right] \tag{4}$$

$$Q_i = \sum_{j=1}^{n_{AC}} V_i V_j \left[G_{ij} \sin\left(\theta_i - \theta_j\right) - B_{ij} \cos\left(\theta_i - \theta_j\right) \right] \tag{5}$$

where $\overline{Y}_{ij} = G_{ij} + jB_{ij}$.

It should be noted that a generator can be either a slack bus or a PV bus, and a load can be a PQ bus with known active and reactive power.

2.2. MT-HVDC Systems Power Flow

The main aim of MT-HVDC systems PF is to determine the DC voltage magnitude at each converter station and the PF within the DC grid. Considering n converter stations in MT-HVDC systems, the following items should be considered to derive the PF equations for the ith converter station of MT-HVDC systems.

- Interface of the converter station with the AC grid
- AC side of the converter station
- Interface of the converter station AC and DC sides
- DC side of MT-HVDC systems
- Control modes of the converter station

2.2.1. Interface of the Converter Station with the AC Grid

The PF equations of at the PCC bus of the ith converter station interfacing with the AC grid is as follows:

$$P_{gi} = \sum_{j=1}^{n_{AC}} V_{gi} V_j \left[G_{ij} \cos\left(\theta_{gi} - \theta_j\right) + B_{ij} \sin\left(\theta_{gi} - \theta_j\right) \right] \tag{6}$$

$$Q_{gi} = \sum_{j=1}^{n_{AC}} V_{gi} V_j \left[G_{ij} \sin\left(\theta_{gi} - \theta_j\right) - B_{ij} \cos\left(\theta_{gi} - \theta_j\right) \right] \tag{7}$$

where $\overline{V}_{gi} = V_{gi} e^{j\theta_{gi}}$.

2.2.2. AC Side of the Converter Station

The PF equations of at the PCC bus of the ith converter station for the asymmetric bipolar MT-HVDC systems is as follows:

$$P_{gi} - P_{gi}^{p} - P_{gi}^{n} = 0 \tag{8}$$

$$Q_{gi} - Q_{gi}^{p} - Q_{gi}^{n} = 0 \tag{9}$$

where P_{gi}^{p}, P_{gi}^{n}, Q_{gi}^{p}, and Q_{gi}^{n} are the active and reactive power of the positive and negative poles of the ith converter station, respectively.

Therefore, the apparent power at the PCC bus of the ith converter station by the positive and negative poles can be written as follows:

$$S_{gi}^{p} = \overline{V}_{gi} \overline{I}_{gi}^{p*} = P_{gi}^{p} + jQ_{gi}^{p} \tag{10}$$

$$S_{gi}^{n} = \overline{V}_{gi} \overline{I}_{gi}^{n*} = P_{gi}^{n} + jQ_{gi}^{n} \tag{11}$$

Applying KCL at the PCC bus of the *i*th converter station leads to deriving the following equation.

$$\bar{I}_{gi} = \bar{I}^p_{gi} + \bar{I}^p_{gi} \tag{12}$$

Using KVL between the PCC bus and the converter terminal of the *i*th converter station leads to deriving the following equations.

$$\bar{I}^p_{gi} = \frac{\overline{V}^p_{ti} - \overline{V}_{gi}}{Z^p_i} = \left(\overline{V}^p_{ti} - \overline{V}_{gi}\right)\left(G^p_i + jB^p_i\right) \tag{13}$$

$$\bar{I}^n_{gi} = \frac{\overline{V}^n_{ti} - \overline{V}_{gi}}{Z^n_i} = \left(\overline{V}^n_{ti} - \overline{V}_{gi}\right)\left(G^n_i + jB^n_i\right) \tag{14}$$

where $Z^p_i = R^p_i + jX^p_i$ and $Z^n_i = R^n_i + jX^n_i$. Also, $\overline{V}^p_{ti} = V^p_{ti}e^{j\theta^p_{ti}}$ and $\overline{V}^n_{ti} = V^n_{ti}e^{j\theta^n_{ti}}$ are the voltage of the positive and negative pole converter terminal, respectively.

Substituting Equations (13) and (14) into Equations (10) and (11), and also separating the active and reactive power, the following equations are obtained.

$$P^p_{gi} = V_{gi}\left[-V_{gi}G^p_i + V^p_{ti}\left\{G^p_i \cos\left(\theta_{gi} - \theta^p_{ti}\right) + B^p_i \sin\left(\theta_{gi} - \theta^p_{ti}\right)\right\}\right] \tag{15}$$

$$Q^p_{gi} = V_{gi}\left[V_{gi}B^p_i + V^p_{ti}\left\{G^p_i \sin\left(\theta_{gi} - \theta^p_{ti}\right) - B^p_i \cos\left(\theta_{gi} - \theta^p_{ti}\right)\right\}\right] \tag{16}$$

$$P^n_{gi} = V_{gi}\left[-V_{gi}G^n_i + V^n_{ti}\left\{G^n_i \cos\left(\theta_{gi} - \theta^n_{ti}\right) + B^n_i \sin\left(\theta_{gi} - \theta^n_{ti}\right)\right\}\right] \tag{17}$$

$$Q^n_{gi} = V_{gi}\left[V_{gi}B^n_i + V^n_{ti}\left\{G^n_i \sin\left(\theta_{gi} - \theta^n_{ti}\right) - B^n_i \cos\left(\theta_{gi} - \theta^n_{ti}\right)\right\}\right] \tag{18}$$

2.2.3. Interface of the Converter Station AC and DC Sides

Regardless of the converter station switching losses, the active power balance between converter station AC and DC sides is as follows:

$$P^p_{ti} = V^p_{DCi}I^p_{DCi} \tag{19}$$

$$P^n_{ti} = V^n_{DCi}I^n_{DCi} \tag{20}$$

where V^p_{DCi}, I^p_{DCi}, V^n_{DCi}, and I^n_{DCi} are the DC bus voltage and current of the positive and negative pole of the *i*th converter station.

Also, the active power at the AC-side terminal of the positive and negative pole of the converter stations can be written as follows:

$$P^p_{ti} = V^p_{ti}\left[V^p_{ti}G^p_i - V_{gi}\left\{G^p_i \cos\left(\theta^p_{ti} - \theta_{gi}\right) + B^p_i \sin\left(\theta^p_{ti} - \theta_{gi}\right)\right\}\right] \tag{21}$$

$$P^n_{ti} = V^n_{ti}\left[V^n_{ti}G^n_i - V_{gi}\left\{G^n_i \cos\left(\theta^n_{ti} - \theta_{gi}\right) + B^n_i \sin\left(\theta^n_{ti} - \theta_{gi}\right)\right\}\right] \tag{22}$$

2.2.4. DC Side of MT-HVDC Systems

When the current flowing through the DC-bus capacitor is zero, $I^p_{DCi} = I^p_{Li}$ and $I^n_{DCi} = I^n_{Li}$, where I^p_{Li} and I^n_{Li} are the current of the positive and negative DC link connected to the *i*th*h* converter station.

The injected current can be determined by the conductance matrix of the DC grid. The DC current injection at the positive and negative poles of the DC bus is as follows:

$$\begin{bmatrix} -I^p_{DC1} \\ \vdots \\ -I^p_{DCi} \\ \vdots \\ -I^p_{DCN} \\ -I^n_{DC1} \\ \vdots \\ -I^n_{DCi} \\ \vdots \\ -I^n_{DCN} \end{bmatrix} = \begin{bmatrix} G_{pp11} & \cdots & G_{pp1i} & \cdots & G_{pp1N} & G_{pn11} & \cdots & G_{pn1i} & \cdots & G_{pn1N} \\ \vdots & \vdots & \vdots & \vdots & \vdots & \vdots & \vdots & \vdots & \vdots & \vdots \\ G_{ppi1} & \cdots & G_{ppii} & \cdots & G_{ppiN} & G_{pni1} & \cdots & G_{pnii} & \cdots & G_{pniN} \\ \vdots & \vdots & \vdots & \vdots & \vdots & \vdots & \vdots & \vdots & \vdots & \vdots \\ G_{ppN1} & \cdots & G_{ppNi} & \cdots & G_{ppNN} & G_{pnN1} & \cdots & G_{pnNi} & \cdots & G_{pnNN} \\ G_{np11} & \cdots & G_{np1i} & \cdots & G_{np1N} & G_{nn11} & \cdots & G_{nn1i} & \cdots & G_{np1N} \\ \vdots & \vdots & \vdots & \vdots & \vdots & \vdots & \vdots & \vdots & \vdots & \vdots \\ G_{npi1} & \cdots & G_{npii} & \cdots & G_{npiN} & G_{nni1} & \cdots & G_{nnii} & \cdots & G_{npiN} \\ \vdots & \vdots & \vdots & \vdots & \vdots & \vdots & \vdots & \vdots & \vdots & \vdots \\ G_{npN1} & \cdots & G_{npNi} & \cdots & G_{npNN} & G_{nnN1} & \cdots & G_{nnNi} & \cdots & G_{npNN} \end{bmatrix} \begin{bmatrix} V^p_{DC1} \\ \vdots \\ V^p_{DCi} \\ \vdots \\ V^p_{DCN} \\ V^n_{DC1} \\ \vdots \\ V^n_{DCi} \\ \vdots \\ V^n_{DCN} \end{bmatrix} \quad (23)$$

Hence, the injected current at the *i*th bus in the DC side is as follows:

$$I^p_{DCi} = -\sum_{j=1}^{N}\left(G_{ppji}V^p_{DCj} + G_{pnji}V^n_{DCj}\right) \quad (24)$$

$$I^n_{DCi} = -\sum_{j=1}^{N}\left(G_{npji}V^p_{DCj} + G_{nnji}V^n_{DCj}\right) \quad (25)$$

where N represents the total number of buses in the DC grid. In addition, G_{ppji}, G_{nnji}, G_{pnji}, and G_{npji} are the conductance of the positive and negative DC link between nodes i and j, respectively.

2.2.5. Control Modes of the Converter Station

There are different modes of operation for each converter station in MT-HVDC systems [4], and based on them, the steady-state equations in the form of equality constraints can be obtained. Also, there are some inequality constraints due to the limits, which are imposed by the converter station voltage and current ratings, as follows:

$$V^{min}_{DCi} \leq V_{DCi} \leq V^{max}_{DCi} \quad (26)$$

$$V^{min}_{i} \leq V_{i} \leq V^{max}_{i} \quad (27)$$

$$I_{gi} \leq I^{max}_{gi} \quad (28)$$

$$I_{DCi} \leq I^{max}_{DCi} \quad (29)$$

3. Mixed AC/DC Power Flow Algorithm

The mixed AC/DC PF algorithm, which is an improved sequential AC/DC PF algorithm [19], can be used to obtain the initial operating points to analyze the dynamics of the hybrid AC/DC grids by solving the AC and DC PF sequentially and keeping both the converter station power and voltage at each node constant.

In order to implement the mixed AC/DC PF algorithm, the per-unit conversion should be performed for the entire system. Each converter station is connected to both AC and DC grids. The AC side of the converter station is modeled by a voltage source connected to the AC bus through a phase reactor ($Z_C = R_C + j\omega L_C$), a capacitor ($Z_F = \frac{-j}{\omega C}$), and a transformer ($Z_{TR} = R_{TR} + j\omega L_{TR}$). Also, the DC side of the converter station is connected to the DC grid. The power losses at the converter station are considered as a quadratic function of the converter station Root Mean Square (RMS) AC current as follows [23]:

$$P_{Loss} = a + bI + cI^2 \tag{30}$$

where a, b, and c are the loss coefficients.

It should be noted that,

$$P_{gi} + P_{DCi} + P_{Lossi} = 0 \tag{31}$$

Assume an MT-HVDC system with n converter stations and n_{DC} DC lines. For AC/DC PF analysis, at least one converter station should be capable of controlling V_{DC} (DC-slack bus) in the entire DC grids, and the rest of the converter stations control the active power. The detailed explanations of the mixed AC/DC PF are given as follows:

1. Start by an initial guess of the active power injected to the AC grid by the DC-slack converter station at the lth iteration ($P_{gn_s}^{(l)}$), where n_s represents the index of the DC-slack bus.
2. Transform all converter stations which are connected to the jth bus to PV or PQ buses based on their control modes (including droop-based control strategies) and solve the AC PF. At the lth iteration, the active power injected by all the non-slack converter stations is constant, while $P_{gn_s}^{(l)}$ changes.
3. Calculate the converter station losses using Equation (30) considering the active power injected by the ith converter to the AC grid ($P_{gi}^{(l)}$) and the active power injected by the ith converter station to the DC grid ($P_{DCi}^{(l)}$). In this step, the AC/DC connections and the converter stations' limits should be considered.
4. Solve the DC PF for the DC grid using the Newton-Raphson method. In this step, the DC-slack bus regulates the DC voltage ($V_{DCn_s}^{(0)}$) initially, and the DC buses determine the active power injection ($P_{DCi}^{(l)}$). Therefore, the DC voltage at each bus and also $P_{DCn_s}^{(l)}$ are calculated.
5. Compute DC-slack and droop buses iteration (k). As a new value of $P_{gn_s}^{(l+1)}$ is calculated, considering the converter station losses,

    I.   Initialization: $P_{gn_s}^{(k=0)} = P_{gn_s}^{(l)}$.

    II. Solve the branch j, $n_s - g$, n_s considering V_{jn_s}, θ_{jn_s}, Q_{jn_s}, and $P_{gn_s}^{(k)}$ using the Newton-Raphson method.

    III. Obtain the new value of $P_{gn_s}^{(k+1)}$ with P_{DCn_s} and P_{Lossn_s} using Equations (30) and (31).

    IV. If $\left| P_{gn_s}^{(k+1)} - P_{gn_s}^{(k)} \right| < \varepsilon$, stop the calculations. Otherwise, $k = k + 1$ and return to step II. The output is $P_{jn_s}^{(k+1)}$.

6. Check the convergence criterion. If $\left| P_{gn_s}^{(k+1)} - P_{gn_s}^{(k)} \right| < \varepsilon$, stop the calculations. Otherwise, $k = k + 1$ and return to step 1.

It should be noted that all linear and nonlinear variables are considered in the mixed AC/DC PF algorithm. Based on the topology of the hybrid AC/DC grids, the rating and length of each transmission line are determined. Also, the droop parameters and reference voltage have a direct impact on the reference power.

4. Results and Discussions

For validation and to demonstrate the performance of the mixed AC/DC PF algorithm, a five-bus AC grid with a three-bus MT-HVDC system is simulated in MATLAB software, as shown in Figure 1. The test system is composed of both AC and DC grids. The data of the system is provided in Tables 1–5.

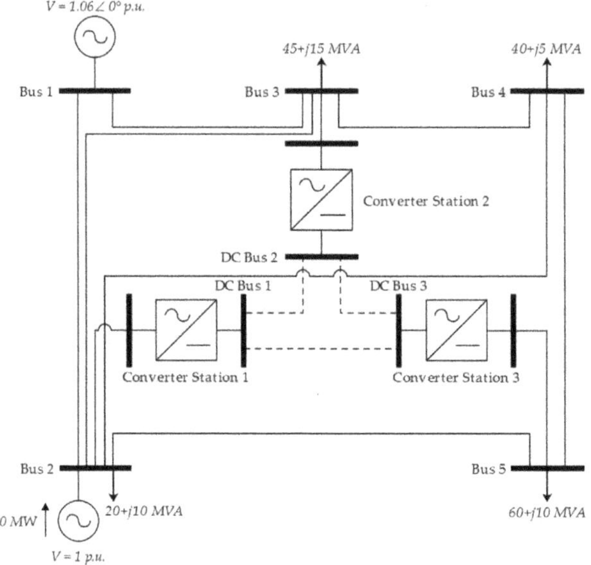

Figure 1. Single-line diagram of a three-bus MT-HVDC system.

Table 1. Parameters of the AC system.

Bus	Type	V (p.u.)	θ (°)	P_G (MW)	Q_G (MVAR)	P_D (MW)	Q_D (MVAR)
1	Slack	1.060	0.00	–	–	0.00	0.00
2	PV	1.000	–	40.00	–	20.00	10.00
3	PQ	–	–	–	–	45.00	15.00
4	PQ	–	–	–	–	40.00	5.00
5	PQ	–	–	–	–	60.00	10.00

Table 2. Parameters of the AC lines.

From	To	R (p.u.)	X (p.u.)	B (p.u.)	S_N (MVA)
1	2	0.02	0.06	0.06	100
1	3	0.08	0.24	0.05	100
2	3	0.06	0.18	0.04	100
2	4	0.06	0.18	0.04	100
2	5	0.04	0.12	0.03	100
3	4	0.01	0.03	0.02	100
4	5	0.08	0.24	0.05	100

Table 3. Parameters of the converter stations in per-unit.

Converter Station	S_N (MVA)	R_{TR}	X_{TR}	B_F	R_{TR}	X_{TR}
1	100	0.0015	0.1121	0.0887	0.0001	0.1643
2	100	0.0015	0.1121	0.0887	0.0001	0.1643
3	100	0.0015	0.1121	0.0887	0.0001	0.1643

Table 4. Power losses coefficients of the converter stations.

Converter Station	a	b	c_{rec}	c_{inv}
1	1.103	0.887	2.885	4.371
2	1.103	0.887	2.885	4.371
3	1.103	0.887	2.885	4.371

Table 5. Parameters of the DC lines.

From	To	R_{DC} (p.u.)	V_{DC} (kV)	P_{DC} (MW)
1	2	0.0260	345	100
1	3	0.0365	345	100
2	3	0.0260	345	100

The parameters of the converter stations for the PF calculations are as follows:

- Converter Station #1: $P - Q$ control mode, $P_g = -60$ MW and $Q_g = -40$ MVAR
- Converter Station #2: $V_{DC} - V$ control mode, $V_{DC} = 1$ p.u. and $V = 1$ p.u.
- Converter Station #3: $P - Q$ control mode, $P_g = 35$ MW and $Q_g = 5$ MVAR

It should be noted that $S_B = 100$ MVA, $V_{DC_B} = 345$ kV, and $V_B = 345$ kV are the base values for the hybrid AC/DC grids per-unit system. Also, the links between the two nodes on the DC grids are bipolar.

The simulations are accomplished using a laptop with the Intel Core i7-8550U processor at 1.80 GHz clock speed and 12-GB of RAM.

4.1. Case 1: AC Power Flow without MT-HVDC Systems

In the first case, the AC PF calculations are performed without considering the DC system. The algorithm is converged in 0.44 s and in three iterations. Tables A1 and A2 show the results of AC PF without DC grids.

4.2. Case 2: AC/DC Power Flow Considering MT-Systems with Constant Active Power and DC Voltage

In the second case, the algorithm is applied to solve the AC/DC PF problem for the studied system considering that converter stations #1 and #3 are operated in constant P-mode and converter station #2 is operated in constant V_{DC}-mode. The algorithm is converged in 0.82 s and in three iterations. Tables A3–A7 show the obtained results of the mixed AC/DC PF calculations in Case 2.

Compared to Case 1, in this case, both the active and reactive PF on each line have decreased and accordingly, the total power losses are decreased. It is observed that the generator connected to bus 2 injects its maximum active power. In addition, compared to the previous case, the total injected reactive power by the two generators is decreased.

4.3. Case 3: AC/DC Power Flow Considering Converter Station Outage

In the third case, the impact of the converter station outage on the AC/DC PF results is analyzed. Tables A8–A11 demonstrate the results of PF calculations both in all AC and DC buses and in branches, in case of the outage of converter station #1. Based on the obtained results, the converter station #3 is capable of operating in constant P-mode. It should be noted that the algorithm is converged in 0.55 s and in three iterations.

In this case, the generator connected to bus 2 injects 40 MW active power to the grid (maximum active power capability) and the total injected reactive power this generator is approximately tripled. The outage of converter station #1 is led to an increase in the active and reactive PF of the AC lines. Therefore, the total power losses on the AC grids are increased. Due to the fact that converter station

#1 is not connected to the grid, the summation of the PF between the DC lines (1-2 and 1-3) is zero and consequently, the overall DC power losses are decreased.

4.4. Case 4: AC/DC Power Flow Considering Droop Control Strategy and Converter Outage

As stated in Section 1.3, the droop control strategy is an efficient way of controlling MT-HVDC systems that can improve the PF of the AC and DC grids after an outage. In the fourth case, it is assumed that all converter stations in the studied system are equipped with the droop controllers. Table A12 illustrates the detailed information of the droop settings for each converter station. Tables A13–A16 show the results of PF calculations both in all AC and DC buses and in branches in case of the outage of converter station #1 and considering droop control settings provided in Table A12. It should be noticed that the algorithm is converged in 0.45 s and in three iterations.

Compared to Case 3, in this case, the contribution of the generator connected to bus 2 to power generation is decreased. As the outage of converter station #1 is considered, compared to Case 3, the PF on the DC lines is decreased. It should be noted that as the performance of each converter station is based on the pre-determined droop parameter, the PF on the DC lines should be in a way that the total power losses on the DC lines become zero.

4.5. Case 5: AC/DC Power Flow Considering Changes of Droop Parameters and Converter Outage

In the fifth case, to demonstrate the impact of changes in the droop parameters of each converter station, the same test as Case 4 is evaluated by changing the droop parameters as shown in Table A17. Tables A18–A21 depict the results of PF calculations both in all AC and DC buses and in branches in case of the outage of converter station #1 and considering the changes in droop parameters according to Table A17. It is worth mentioning that the algorithm is converged in 0.43 s and in three iterations.

Changing the droop parameters is led to an increase in the total active power generated by the generator connected to bus 2. As a consequence, there are some slight changes of the PF on the AC lines. However, compared to the previous case, those changes are negligible. In addition, changing the droop parameters is causes an increase in the DC PF on the DC lines (compared to Case 4), but as the converter station #1 is not connected to the grid, the summation of the PF between the DC lines is equal to zero.

4.6. Case 6: AC/DC Power Flow Considering Converter Station Limits and Converter Outage

In the sixth case, the impact of the converter station limits on the AC/DC PF solution is investigated. When an active power set-point of a converter station equipped with the P controller is outside of the P-Q capability chart (P-Q capability chart shows the possible operation points.), the active power order should be reduced to comply with the predefined limit. Similarly, when a reactive power set-point of a converter station equipped with the Q controller is outside of the P-Q capability chart, the reactive power order should be reduced to comply with the predefined limit, subject to not reaching the active power limit. In addition, when a converter station equipped with the V_{DC}-droop controller reaches its limit, the converter station should be set to a constant P injection equal to the maximum active power limit of the converter station. In addition, when a converter station equipped with the V-droop controller reaches its limit, the converter station should be set to a constant Q injection based on the predefined limit. For both converter stations equipped with the Q controller and V_{DC}-droop controller, the priority is given to active power over reactive power, when enforcing the limits. It should be noted that however, all the DC-slack buses are disregarded from the analysis, they are rechecked at the end of AC/DC PF calculations.

According to the above explanations, it is assumed that the converter station #1 reaches the reactive power limit and the converter station control is changed from constant V-mode to constant Q-mode. Tables A22–A25 illustrate the results of PF calculations in all both AC and DC buses and branches after enforcing the converter station's current and voltage limits. Meanwhile, the algorithm is converged in 0.60 s and in three iterations. After three iterations, the reactive power set-point of the converter station #1 is changed from −40 MVAR to −36.25 MVAR.

To demonstrate the impact of changing the set-point of the converter station, the active power set-point of the converter station #1 is set to −130 MW so that simultaneously both active and reactive power violate their limits. Tables A26–A29 show the results of PF calculations both in all AC and DC buses and in branches after changing the active power set-point of the converter station #1 to −130 MW. It should be noted that the algorithm is converged in 0.71 s and in three iterations.

After three iterations, the active and reactive power set-points of the converter station #1 are changed from −130 MW to −121.21 MW, and −40 MVAR to −8.96 MVAR, respectively. In this case, the total active power generated by the generation units is increased (tangible changes are related to the one connected to bus 2). The overall power losses on the AC lines are decreased. As the converter station #1 is connected to the grid and its set-points are reached their maximums limits, the active power losses (correspond to ZI^2) are increased. Therefore, the PF on the DC lines is increased.

4.7. Case 7: AC/DC Power Flow for the Large-Scale Hybrid AC/DC Grids with MT-HVDC Systems

In order to validate and show the performance of the mixed AC/DC PF algorithm for the large-scale hybrid AC/DC grids with the integration of MT-HVDC system, the modified IEEE 39-bus test system with two four-bus MT-HVDC systems (with different colors) is simulated in MATLAB software, as shown in Figure 2. The data of the MT-HVDC systems are provided in Tables A30–A32.

As it can be observed from Figure 2, the IEEE 39-bus test system is divided into three areas, in which area 1 comprises of three coherent generators (G8, G9, and G10), area 2 comprises of three coherent generators (G1, G2, and G3), and area 3 comprises of four coherent generators (G4, G5, G6, and G7). The main corridors to interconnect those three areas are given as follows:

- Area 1 to Area 2: Lines from bus 1 to bus 39, and from bus 3 to bus 4
- Area 1 to Area 3: Lines from bus 3 to bus 18, and from bus 27 to bus 17
- Area 2 to Area 3: Line from bus 14 to bus 15

The parameters of the converter stations for the PF calculations are as follows:

- Converter Station #1: $P - Q$ control mode, $P_g = -60$ MW and $Q_g = -20$ MVAR
- Converter Station #2: $V_{DC} - V$ control mode, $V_{DC} = 1$ p.u. and $V = 1$ p.u.
- Converter Station #3: $P - Q$ control mode, $P_g = 40$ MW and $Q_g = 40$ MVAR
- Converter Station #4: $P - Q$ control mode, $P_g = 40$ MW and $Q_g = 40$ MVAR
- Converter Station #5: $V_{DC} - V$ control mode, $V_{DC} = 1$ p.u. and $V = 1$ p.u.
- Converter Station #6: $P - Q$ control mode, $P_g = 40$ MW and $Q_g = 30$ MVAR
- Converter Station #7: $P - Q$ control mode, $P_g = 40$ MW and $Q_g = 20$ MVAR
- Converter Station #8: $P - Q$ control mode, $P_g = 40$ MW and $Q_g = 40$ MVAR

From the operation and planning perspectives, the outage of the mentioned lines can cause entirely disconnection of the two areas from each other and accordingly, prevention of power exchange between two areas, operating in islanded mode, and instability of the AC grids. Therefore, there is a need for strengthening the power transmission lines among the mentioned areas. To do so, two four-bus MT-HVDC systems are considered to interconnect the mentioned areas together. Areas 1 and 3 are interconnected with four converter stations (CS1, CS2, CS3, and CS4) and areas 2 and 3 are interconnected with four converter stations (CS5, CS6, CS7, and CS8). Hence, in case of AC lines outage, the power can be transferred via DC links. Tables A33 and A34 show the results of AC PF on the IEEE 39-bus test system without DC grids (Case 7-1).

Tables A35–A38 show the obtained results of the mixed AC/DC PF calculations when MT-HVDC systems are connected to the IEEE 39-bus test system (Case 7-2). In a general view, due to the fact that by interconnecting the MT-HVDC systems to the grid, more power is required, the total generated active and reactive power by the generators should increase and decrease, respectively. There is no connection between the MT-HVDC systems and any of the generators. Therefore, only the voltage magnitude and angles are changed. This should be considered that each converter has its own power losses. Therefore, compared to the AC PF, the total power losses on the AC lines in the mixed AC/DC PF is increased by 0.11 MW. There are some slight changes in the total active power generated by the generator #31. However, the generated reactive power by all the generators is changed. The changes of the PF on the AC lines 1–39, 3–4, 3–18, 4–14, 14–15, 15–16, and 17–27 are significant since they are directly connected to the MT-HVDC systems. To check the changes of the PF on the other lines, the connected lines to the mentioned lines are checked and it is observed that there are some slight changes of the PF on the lines 4-5 and 13-14. Based on the obtained results, the slack converters are capable of transferring power on the DC lines based on the total demand. It is also noticed that compared to the other DC lines, the changes of the PF on the lines 5–6, 5–7, and 5–8 are considerable and the changes of the PF on the DC lines 6–7 and 6–8 are almost zero.

To minimize the power losses, in the next case, it is considered that the converters are equipped with the droop controllers. Table A39 illustrates the detailed information of the droop settings for each converter station. Tables A40–A43 illustrate the results of mixed AC/DC PF considering droop control settings provided in Table A39 (Case 7-3). Analyzing the obtained results shows that the changes of the PF on the AC lines 1–39, 2–3, and 17–27 are almost negligible but the total power losses on the AC lines are decreased. The DC lines 1–4 and 2–3 have the most power losses. In addition, no power is transferred on lines 6–7 and 6–8. Therefore, they can be considered as reserve lines for planning purposes. It is worth mentioning that compared to Case 7-2, the total power losses on the DC lines are decreased.

To analyze the impact of MT–HVDC systems on the PF analysis, AC lines 3–4, 3–18, 4–14, and 14–15 as the main interconnected corridors between each of area are disconnected. Tables A44 and A45 show the results of AC PF without DC grids in the case of disconnecting the AC lines 3–4, 3–18, 4–14, and 14–15 (Case 7-4). The disconnection of the mentioned lines is caused that some of the lines reach their maximum transfer power capabilities. The changes of the PF on the AC lines 4–5, 5–6, 6–11, and 10–11 are drastically changed. However, the changes of the PF on the AC lines 10–13 and 13–14 are deceased. The rest of the lines have either no or very slight changes on the PF.

To improve the reliability of the hybrid AC/DC grids, the MT-HVDC systems are connected to compensate for the disconnections of the AC lines 3–4, 3–18, 4–14, and 14–15. Tables A46–A49 show the obtained results of the mixed AC/DC PF calculations when MT-HVDC systems are connected to the IEEE 39-bus test system considering the disconnections of the AC lines 3–4, 3–18, 4–14, and 14–15 (Case 7-5). Based on the obtained results, the significant changes on the PF are related to the AC lines 1–2, 1–39, 10–13, 13–14, and 17–27 (increase) and 2–25, 5–6, 6–11, 10–11, 15–16, 17–18, and 26–27 (decrease). The total power losses on the AC lines are closed to the previous case. In addition, DC lines 5–6, 5–7, and 5–8 are the DC lines with high power losses.

In order to minimize the power losses, it is assumed that the converter stations are equipped with the droop controllers as Table A39. Tables A50–A53 show the obtained results of the mixed AC/DC PF calculations when MT-HVDC systems are connected to the IEEE 39-bus test system considering the droop parameters and disconnections of the AC lines 3–4, 3–18, 4–14, and 14–15 (Case 7-6). The obtained results show that compared to the previous case, only the PF of the AC lines 4–5, 5–6, 15–16, 17–18, and 26–27 are increased and the total power losses both on the AC and DC lines are decreased.

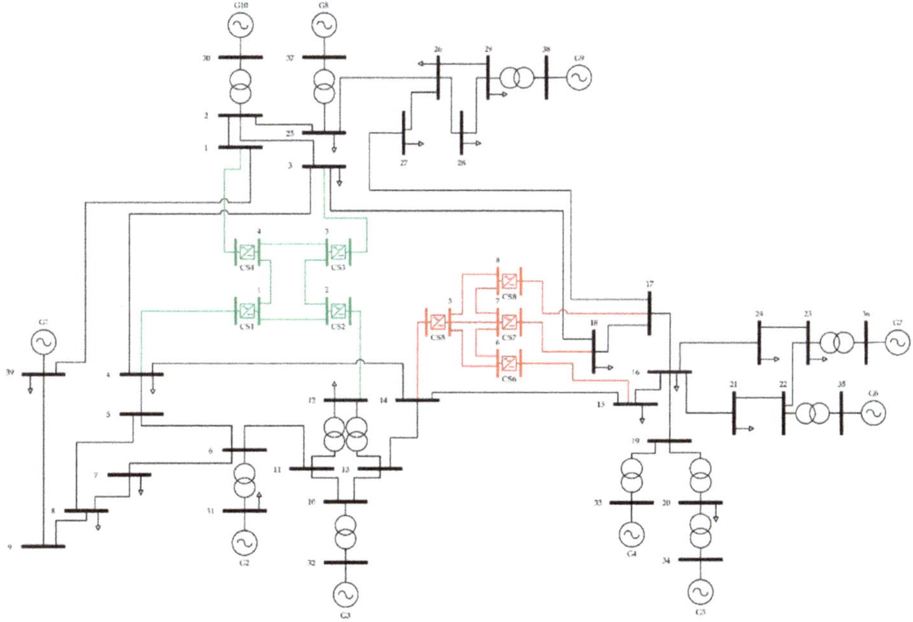

Figure 2. Single-line diagram of the modified IEEE 39-bus test system with two four-bus MT-HVDC systems.

5. Conclusions

In this paper, a mixed AC/DC Power Flow (PF) algorithm for the steady-state interaction of the large-scale MT-HVDC systems is investigated. This algorithm is an improved sequential AC/DC PF algorithm, which uses the Newton-Raphson method to solve the DC PF problem. Different operational constraints and control strategies along with contingency analysis in the hybrid AC/DC grids are considered in this study. Fast convergence and high accuracy are the main advantages of the mixed AC/DC PF algorithm. In addition, it is a powerful tool for sensitivity analysis and congestion management in power systems. Various cases are studied in this paper to evaluate the performance of the mixed AC/DC PF algorithm. The obtained results demonstrate the robustness and effectiveness of the mixed AC/DC PF algorithm for power system operation and planning studies.

Author Contributions: F.M. was responsible for methodology, collecting resources, data analysis, writing—original draft preparation, and writing—review and editing. G.-A.N. and M.S. were responsible for the supervision, and writing—review and editing. All authors have read and agreed to the published version of the manuscript.

Funding: M. Saif acknowledges the financial support of Natural Sciences and Engineering Research Council (NSERC) of Canada, as well as National Science Foundation of China under grant number 61873144.

Appendix A

Note: P and Q are in MW and $MVAR$, respectively.

Table A1. Results of AC PF calculations in Case 1. AC bus data.

Bus	Voltage		Generation		Load	
	$\|V\|$ (p.u.)	θ (°)	P	Q	P	Q
1	1.060	0.000	131.12	90.82	–	–
2	1.000	−2.061	40.00	−61.59	20.00	10.00
3	0.987	−4.637	–	–	45.00	15.00
4	0.984	−4.957	–	–	40.00	5.00
5	0.972	−5.765	–	–	60.00	10.00
Total:			171.12	29.23	165.00	40.00

Table A2. Results of AC PF calculations in Case 1. AC branch data.

Branch	From	To	From Bus Injection		To Bus Injection		Power Losses	
			P	Q	P	Q	P	Q
1	1	2	89.33	74.00	−86.85	−72.91	2.486	7.46
2	1	3	41.79	16.82	−40.27	−17.51	1.518	4.55
3	2	3	24.47	−2.52	−24.11	−0.35	0.360	1.08
4	2	4	27.71	−1.72	−27.25	−0.83	0.461	1.38
5	2	5	54.66	5.56	−53.44	−4.83	1.215	3.65
6	3	4	19.39	2.86	−19.35	−4.69	0.040	0.12
7	4	5	6.60	0.52	−6.56	−5.17	0.043	0.13
						Total:	6.123	18.37

Table A3. Results of the mixed AC/DC PF calculations in Case 2. AC bus data.

Bus	Voltage		Generation		Load	
	$\|V\|$ (p.u.)	θ (°)	P	Q	P	Q
1	1.060	0.000	133.64	84.32	–	–
2	1.000	−2.383	40.00	−32.84	20.00	10.00
3	1.000	−3.895	–	–	45.00	15.00
4	0.996	−4.262	–	–	40.00	5.00
5	0.991	−4.149	–	–	60.00	10.00
Total:			173.64	51.48	165.00	40.00

Table A4. Results of the mixed AC/DC PF calculations in Case 2. AC branch data.

Branch	From	To	From Bus Injection		To Bus Injection		Power Losses	
			P	Q	P	Q	P	Q
1	1	2	98.38	71.37	−95.66	−69.59	2.717	8.15
2	1	3	35.26	12.96	−34.20	−15.08	1.062	3.19
3	2	3	13.25	−6.22	−13.14	2.57	0.116	0.35
4	2	4	17.08	−5.18	−16.89	1.74	0.181	0.54
5	2	5	25.33	−1.85	−25.07	−0.35	0.257	0.77
6	3	4	23.09	4.64	−23.04	−6.47	0.057	0.17
7	4	5	−0.07	−0.27	0.07	−4.65	0.004	0.01
						Total:	4.394	13.18

Table A5. Results of the mixed AC/DC PF calculations in Case 2. DC bus data.

Bus DC	Bus AC	Voltage Magnitude (p.u.)	Active Power (MW)
1	2	1.008	−58.627
2	3	1.000	21.901
3	5	0.998	36.186

Table A6. Results of the mixed AC/DC PF calculations in Case 2. Converter station data.

Bus DC	Bus Injection		Converter Voltage		Total Loss
	P	Q	$|V|$ (p.u.)	θ (°)	P
1	−60.00	−40.00	0.890	−13.017	1.37
2	20.76	7.14	1.007	−0.655	1.14
3	35.00	5.00	0.995	1.442	1.19
				Total:	3.70

Bus DC	Converter Power		Filter	Transformer Loss		Reactor Loss		Converter Loss
	P	Q	Q	P	Q	P	Q	P
1	−59.22	−32.63	−8.12	0.08	5.83	0.01	9.66	1.29
2	20.76	−0.65	−9.02	0.01	0.54	0.00	0.70	1.14
3	35.02	−0.37	−8.83	0.02	1.43	0.00	2.03	1.17
			Total:	0.11	7.80	0.01	12.39	3.60

Bus DC	Grid Power		Trans. Filter Power		Filter	Conv. Filter Power	Conv. Power	
	P	Q	P	Q	Q	Q	P	Q
1	−60.00	−40.00	−59.92	−34.17	−8.12	−42.29	−59.92	−32.63
2	20.76	7.14	20.76	7.68	−9.02	−1.35	20.76	−0.65
3	35.00	5.00	35.02	6.43	−8.83	−2.40	35.02	−0.37

Table A7. Results of the mixed AC/DC PF calculations in Case 2. DC branch data.

Branch			From Bus	To Bus	Power Losses
	From	To	P	P	P
1	1	2	30.66	−30.42	0.24
2	2	3	8.52	−8.50	0.02
3	1	3	27.96	−27.68	0.28
				Total:	0.54

Table A8. Results of the mixed AC/DC PF calculations in Case 3. AC bus data.

Bus	Generation		Load	
	P	Q	P	Q
1	133.93	84.93	–	–
2	40.00	−90.48	20.00	10.00
3	–	–	45.00	15.00
4	–	–	40.00	5.00
5	–	–	60.00	10.00
Total:	173.93	−5.55	165.00	40.00

Table A9. Results of the mixed AC/DC PF calculations in Case 3. AC branch data.

Branch			From Bus Injection		To Bus Injection		Power Losses	
	From	To	P	Q	P	Q	P	Q
1	1	2	84.95	75.29	−82.56	−74.50	2.386	7.16
2	1	3	48.98	9.64	−47.17	−9.50	1.819	5.46
3	2	3	34.58	−12.23	−33.80	10.57	0.780	2.34
4	2	4	34.13	−9.79	−33.40	8.01	0.735	2.21
5	2	5	33.85	−3.96	−33.39	2.38	0.461	1.38
6	3	4	−1.69	13.76	1.71	−15.69	0.022	0.07
7	4	5	−8.31	2.68	8.39	−7.38	0.077	0.23
						Total:	6.283	18.85

Table A10. Results of the mixed AC/DC PF calculations in Case 3. DC bus data.

Bus DC	Bus Injection		Total Loss
	P	Q	P
1	0.00	0.00	0.00
2	−37.65	29.84	1.22
3	35.00	5.00	1.19
		Total:	2.41

Table A11. Results of the mixed AC/DC PF calculations in Case 3. DC branch data.

Branch			From Bus	To Bus	Power Losses
	From	To	P	P	P
1	1	2	−10.67	10.70	0.03
2	2	3	25.73	−25.55	0.17
3	1	3	10.67	−10.63	0.04
				Total:	0.24

Table A12. Droop control settings for each converter station.

Converter Station	Droop Parameter	P^*_{DC} *(MW)*	V^*_{DC} *(p.u.)*
1	0.005	−58.6274	1.0079
2	0.007	21.9013	1.0000
3	0.005	36.1856	0.9778

* P^*_{DC} and V^*_{DC} are the reference power and reference DC voltage, respectively.

Table A13. Results of the mixed AC/DC PF calculations in Case 4. AC bus data.

Bus	Generation		Load	
	P	Q	P	Q
1	133.39	84.72	–	–
2	40.00	−81.05	20.00	10.00
3	–	–	45.00	15.00
4	–	–	40.00	5.00
5	–	–	60.00	10.00
Total:	173.39	0.67	165.00	40.00

Table A14. Results of the mixed AC/DC PF calculations in Case 4. AC branch data.

Branch			From Bus Injection		To Bus Injection		Power Losses	
	From	To	P	Q	P	Q	P	Q
1	1	2	90.34	73.70	−87.83	−72.54	2.510	7.53
2	1	3	43.05	11.02	−41.59	−11.96	1.456	4.37
3	2	3	25.47	−9.78	−25.05	7.06	0.426	1.28
4	2	4	28.39	−7.76	−27.88	5.29	0.503	1.51
5	2	5	53.97	−0.97	−52.81	1.52	1.165	3.50
6	3	4	18.07	10.21	−18.03	−12.07	0.045	0.14
7	4	5	5.91	1.77	−5.87	−6.52	0.043	0.13
						Total:	6.148	18.46

Table A15. Results of the mixed AC/DC PF calculations in Case 4. DC bus data.

Bus DC	Bus Injection		Total Loss
	P	Q	P
1	0.00	0.00	0.00
2	−3.57	20.31	1.13
3	1.33	5.00	1.11
Total:			2.24

Table A16. Results of the mixed AC/DC PF calculations in Case 4. DC branch data.

Branch	From	To	From Bus P	To Bus P	Power Losses P
1	1	2	−0.72	0.72	0.00
2	2	3	1.72	−1.72	0.00
3	1	3	0.72	−0.72	0.00
				Total:	0.00

Table A17. Changes in the droop parameters of each converter station.

Converter Station	Droop Parameter
1	0.0010
2	0.0014
3	0.0010

Table A18. Results of the mixed AC/DC PF calculations in Case 5. AC bus data.

Bus	Generation		Load	
	P	Q	P	Q
1	133.38	84.72	–	–
2	40.00	−81.22	20.00	10.00
3	–	–	45.00	15.00
4	–	–	40.00	5.00
5	–	–	60.00	10.00
Total:	173.38	3.50	165.00	40.00

Table A19. Results of the mixed AC/DC PF calculations in Case 5. AC branch data.

Branch	From	To	From Bus Injection P	From Bus Injection Q	To Bus Injection P	To Bus Injection Q	Power Losses P	Power Losses Q
1	1	2	90.24	73.73	−87.74	−72.58	2.508	7.52
2	1	3	43.14	11.00	−41.68	−11.92	1.461	4.38
3	2	3	25.62	−9.82	−25.19	7.11	0.430	1.29
4	2	4	28.48	−7.80	−27.97	5.34	0.507	1.52
5	2	5	53.64	−1.02	−52.49	1.54	1.151	3.45
6	3	4	17.75	10.26	−17.71	−12.12	0.044	0.13
7	4	5	5.68	1.78	−5.64	−6.54	0.041	0.12
						Total:	6.142	18.41

Table A20. Results of the mixed AC/DC PF calculations in Case 5. DC bus data.

Bus DC	Bus Injection		Total Loss
	P	*Q*	*P*
1	0.00	0.00	0.00
2	−4.11	20.45	1.13
3	1.87	5.00	1.11
Total:			2.24

Table A21. Results of the mixed AC/DC PF calculations in Case 5. DC branch data.

Branch			From Bus	To Bus	Power Losses
	From	To	*P*	*P*	*P*
1	1	2	−0.88	0.88	0.00
2	2	3	2.11	−2.11	0.00
3	1	3	0.88	−0.88	0.00
				Total:	0.00

Table A22. Results of the mixed AC/DC PF calculations in Case 6-1. AC bus data.

Bus	Generation		Load	
	P	*Q*	*P*	*Q*
1	133.38	84.72	–	–
2	40.00	−81.22	20.00	10.00
3	–	–	45.00	15.00
4	–	–	40.00	5.00
5	–	–	60.00	10.00
Total:	173.38	3.50	165.00	40.00

Table A23. Results of the mixed AC/DC PF calculations in Case 6-1. AC branch data.

Branch			From Bus Injection		To Bus Injection		Power Losses	
	From	To	*P*	*Q*	*P*	*Q*	*P*	*Q*
1	1	2	90.24	73.73	−87.74	−72.58	2.508	7.52
2	1	3	43.14	11.00	−41.68	−11.92	1.461	4.38
3	2	3	25.62	−9.82	−25.19	7.11	0.430	1.29
4	2	4	28.48	−7.80	−27.97	5.34	0.507	1.52
5	2	5	53.64	−1.02	−52.49	1.54	1.151	3.45
6	3	4	17.75	10.26	−17.71	−12.12	0.044	0.13
7	4	5	5.68	1.78	−5.64	−6.54	0.041	0.12
						Total:	6.142	18.41

Table A24. Results of the mixed AC/DC PF calculations in Case 6-1. DC bus data.

Bus DC	Bus Injection		Total Loss
	P	*Q*	*P*
1	0.00	0.00	0.00
2	−4.11	20.45	1.13
3	1.87	5.00	1.11
Total:			2.24

Table A25. Results of the mixed AC/DC PF calculations in Case 6-1. DC branch data.

Branch	From	To	From Bus P	To Bus P	Power Losses P
1	1	2	−0.88	0.88	0.00
2	2	3	2.11	−2.11	0.00
3	1	3	0.88	−0.88	0.00
				Total:	0.00

Table A26. Results of the mixed AC/DC PF calculations in Case 6-2. AC bus data.

Bus	Generation P	Q	Load P	Q
1	135.63	83.69	–	–
2	40.00	−61.38	20.00	10.00
3	–	–	45.00	15.00
4	–	–	40.00	5.00
5	–	–	60.00	10.00
Total:	175.63	22.31	165.00	40.00

Table A27. Results of the mixed AC/DC PF calculations in Case 6-2. AC branch data.

Branch	From	To	From Bus Injection P	Q	To Bus Injection P	Q	Power Losses P	Q
1	1	2	113.59	67.07	−110.41	−63.90	3.180	9.54
2	1	3	22.04	16.62	−21.42	−20.08	0.615	1.84
3	2	3	−7.87	0.69	7.91	−4.57	0.042	0.12
4	2	4	0.18	0.16	−0.18	−4.13	0.003	0.01
5	2	5	16.89	0.63	−16.78	−3.25	0.116	0.35
6	3	4	48.34	−3.45	−48.10	2.16	0.234	0.70
7	4	5	8.28	−3.02	−8.22	−1.75	0.056	0.17
						Total:	4.246	12.73

Table A28. Results of the mixed AC/DC PF calculations in Case 6-2. DC bus data.

Bus DC	Bus Injection P	Q	Total Loss P
1	−121.21	8.96	1.69
2	79.82	−13.10	1.38
3	35.00	5.00	1.19
Total:			4.26

Table A29. Results of the mixed AC/DC PF calculations in Case 6-2. DC branch data.

Branch	From	To	From Bus P	To Bus P	Power Losses P
1	1	2	73.60	−72.24	1.36
2	2	3	−8.97	8.99	0.02
3	1	3	45.92	−45.18	0.74
				Total:	2.12

Table A30. Parameters of the converter stations in per-unit.

Converter Station	S_N (MVA)	R_{TR}	X_{TR}	B_F	R_{TR}	X_{TR}
1	100	0.0015	0.1121	0.0887	0.0001	0.1643
2	100	0.0015	0.1121	0.0887	0.0001	0.1643
3	100	0.0015	0.1121	0.0887	0.0001	0.1643
4	100	0.0015	0.1121	0.0887	0.0001	0.1643
5	100	0.0015	0.1121	0.0887	0.0001	0.1643
6	100	0.0015	0.1121	0.0887	0.0001	0.1643
7	100	0.0015	0.1121	0.0887	0.0001	0.1643

Table A31. Power losses coefficients of the converter stations.

Converter Station	a	b	c_{rec}	c_{inv}
1	1.103	0.887	2.885	4.371
2	1.103	0.887	2.885	4.371
3	1.103	0.887	2.885	4.371
4	1.103	0.887	2.885	4.371
5	1.103	0.887	2.885	4.371
6	1.103	0.887	2.885	4.371
7	1.103	0.887	2.885	4.371

Table A32. Parameters of the DC lines.

From	To	R_{DC} (p.u.)	V_{DC} (kV)	P_{DC} (MW)
1	2	0.0520	345	100
1	4	0.0520	345	100
2	3	0.0520	345	100
3	4	0.0520	345	100
5	6	0.0730	345	100
5	7	0.0730	345	100
5	8	0.0730	345	100
6	7	0.0730	345	100
6	8	0.0730	345	100

Table A33. Results of AC PF calculations in Case 7-1. AC bus data.

Bus	Voltage		Generation		Load	
	$\|V\|$ (p.u.)	θ (°)	P	Q	P	Q
1	1.039	−13.537	–	–	97.60	44.20
2	1.048	−9.785	–	–	–	–
3	1.031	−12.276	–	–	322.00	2.40
4	1.004	−12.627	–	–	500.00	184.00
5	1.006	−11.192	–	–	–	–
6	1.008	−10.408	–	–	–	–
7	0.998	−12.756	–	–	233.80	84.00
8	0.998	−13.336	–	–	522.00	176.60
9	1.038	−14.178	–	–	6.50	−66.60
10	1.018	−8.171	–	–	–	–
11	1.013	−8.937	–	–	–	–
12	1.001	−8.999	–	–	8.53	88.00
13	1.015	−8.930	–	–	–	–
14	1.012	−10.715	–	–	–	–
15	1.016	−11.345	–	–	320.00	153.00
16	1.033	−10.033	–	–	329.00	32.30
17	1.034	−11.116	–	–	–	–
18	1.032	−11.986	–	–	158.00	30.00
19	1.050	−5.410	–	–	–	–
20	0.991	−6.821	–	–	680.00	103.00
21	1.032	−7.629	–	–	274.00	115.00
22	1.050	−3.183	–	–	–	–
23	1.045	−3.381	–	–	247.50	84.60
24	1.038	−9.914	–	–	308.60	−92.20
25	1.058	−8.369	–	–	224.00	47.20
26	1.053	−9.439	–	–	139.00	17.00
27	1.038	−11.362	–	–	281.00	75.50
28	1.050	−5.928	–	–	206.00	27.60
29	1.050	−3.170	–	–	283.50	26.90
30	1.050	−7.370	250.00	161.76	–	–
31	0.982	0.000	677.87	221.57	9.20	4.60
32	0.984	−0.188	650.00	206.96	–	–
33	0.997	−0.196	632.00	108.29	–	–
34	1.012	−1.631	508.00	166.69	–	–
35	1.049	1.777	650.00	210.66	–	–
36	1.064	4.468	560.00	100.16	–	–
37	1.028	−1.583	540.00	−1.37	–	–
38	1.027	3.893	830.00	21.73	–	–
39	1.030	−14.535	1000.0	78.47	1104.00	250.00
	Total:		6297.87	1274.92	6254.23	1387.10

Table A34. Results of AC PF calculations in Case 7-1. AC branch data.

Branch	From	To	From Bus Injection		To Bus Injection		Power Losses	
			P	Q	P	Q	P	Q
1	1	2	−173.7	−40.31	174.68	−24.36	0.978	11.48
2	1	39	76.10	−3.89	−76.03	−74.75	0.066	1.65
3	2	3	319.91	88.59	−318.58	−100.88	1.335	15.51
4	2	25	−244.59	82.97	248.93	−93.84	4.337	5.33
5	2	30	−250.00	−147.20	250.00	161.76	0.000	14.56
6	3	4	37.34	113.06	−37.13	−132.59	0.208	3.40
7	3	18	−40.76	−14.59	40.78	−7.94	0.017	0.21
8	4	5	−197.45	−4.09	197.76	−4.52	0.309	4.95
9	4	14	−265.42	−47.32	265.99	42.48	0.571	9.22
10	5	6	−536.94	−43.11	537.51	46.16	0.573	7.45
11	5	8	339.18	47.64	−338.24	−49.39	0.933	13.07
12	6	7	453.82	81.55	−452.56	−73.59	1.261	19.33
13	6	11	−322.65	−38.85	323.38	33.14	0.724	8.48
14	6	31	−668.67	−88.85	668.67	216.97	0.000	128.12
15	7	8	218.76	−10.41	−218.56	4.84	0.192	2.21
16	8	9	34.81	−132.06	−34.48	97.72	0.324	5.11
17	9	39	27.98	−31.12	−27.97	−96.78	0.018	0.44
18	10	11	327.90	73.37	−327.46	−76.18	0.438	4.71
19	10	13	322.10	37.49	−321.69	−40.65	0.407	4.38
20	10	32	−650.00	−110.87	650.00	206.96	0.000	96.1
21	12	11	−4.06	−42.25	4.09	43.04	0.029	0.79
22	12	13	−4.47	−45.75	4.51	46.68	0.034	0.93
23	13	14	317.18	−6.03	−316.30	−1.80	0.879	9.87
24	14	15	50.31	−40.68	−50.26	3.66	0.053	0.64
25	15	16	−269.74	−156.66	270.56	147.33	0.825	8.61
26	16	17	224.02	−42.54	−223.68	32.50	0.338	4.29
27	16	19	−451.30	−54.20	454.38	58.75	3.078	37.52
28	16	21	−329.60	14.44	330.42	−27.74	0.821	13.86
29	16	24	−42.68	−97.33	42.71	90.63	0.030	0.59
30	17	18	199.04	11.05	−198.78	−22.06	0.261	3.06
31	17	27	24.64	−43.56	−24.62	9.23	0.016	0.21
32	19	20	174.73	−9.170	−174.51	13.48	0.218	4.30
33	19	33	−629.11	−49.58	632.00	108.29	2.894	58.71
34	20	34	−505.49	−116.48	508.00	166.69	2.511	50.21
35	21	22	−604.42	−87.26	607.21	108.15	2.783	48.7
36	22	23	42.790	41.88	−42.77	−61.75	0.025	0.40
37	22	35	−650.00	−150.04	650.00	210.66	0.000	60.63
38	23	24	353.84	−0.50	−351.31	1.57	2.529	40.24
39	23	36	−558.57	−22.35	560.00	100.16	1.430	77.82
40	25	26	65.41	−18.81	−65.29	−39.04	0.126	1.27
41	25	37	−538.34	65.45	540.00	−1.37	1.657	64.08
42	26	27	257.30	68.21	−256.38	−84.73	0.920	9.66
43	26	28	−140.82	−21.21	141.61	−56.36	0.788	8.69
44	26	29	−190.19	−24.96	192.10	−67.79	1.914	20.98
45	28	29	−347.61	28.76	349.16	−39.44	1.556	16.78
46	29	38	−824.77	80.33	830.0	21.73	5.234	102.06
						Total:	43.640	1000.61

Table A35. Results of Mixed AC/DC PF calculations in Case 7-2. AC bus data.

Bus	Voltage		Generation		Load	
	$\|V\|$ (p.u.)	θ (°)	P	Q	P	Q
1	1.047	−12.439	–	–	97.60	44.20
2	1.051	−8.959	–	–	–	–
3	1.034	−11.548	–	–	322.00	2.400
4	0.998	−12.879	–	–	500.00	184.00
5	1.001	−11.443	–	–	–	–
6	1.004	−10.675	–	–	–	–
7	0.994	−12.969	–	–	233.80	84.00
8	0.994	−13.517	–	–	522.00	176.60
9	1.037	−13.804	–	–	6.50	−66.60
10	1.012	−8.658	–	–	–	–
11	1.009	−9.390	–	–	–	–
12	1.000	−10.075	–	–	8.530	88.00
13	1.008	−9.464	–	–	–	–
14	1.000	−11.243	–	–	–	–
15	1.016	−10.879	–	–	320.00	153.00
16	1.035	−9.343	–	–	329.00	32.30
17	1.040	−10.213	–	–	–	–
18	1.038	−11.041	–	–	158.00	30.00
19	1.051	−4.727	–	–	–	–
20	0.991	−6.135	–	–	680.00	103.00
21	1.034	−6.945	–	–	274.00	115.00
22	1.051	−2.507	–	–	–	–
23	1.046	−2.705	–	–	247.50	84.60
24	1.040	−9.223	–	–	308.60	−92.20
25	1.060	−7.519	–	–	224.00	47.20
26	1.056	−8.561	–	–	139.00	17.00
27	1.043	−10.465	–	–	281.00	75.50
28	1.052	−5.060	–	–	206.00	27.60
29	1.051	−2.306	–	–	283.50	26.90
30	1.050	−6.550	250.00	147.24	–	–
31	0.982	0.000	691.88	240.39	9.20	4.600
32	0.984	−0.633	650.00	232.33	–	–
33	0.997	0.488	632.00	103.03	–	–
34	1.012	−0.946	508.00	164.25	–	–
35	1.049	2.449	650.00	204.75	–	–
36	1.064	5.139	560.00	96.81	–	–
37	1.028	−0.744	540.00	−10.62	–	–
38	1.026	4.752	830.00	14.76	–	–
39	1.030	−13.788	1000.0	54.67	1104.00	250.00
Total:			6311.88	1247.61	6254.23	1387.10

Table A36. Results of Mixed AC/DC PF calculations in Case 7-2. AC branch data.

Branch	From	To	From Bus Injection		To Bus Injection		Power Losses	
			P	Q	P	Q	P	Q
1	1	2	−161.82	−30.43	162.66	−36.59	0.838	9.85
2	1	39	104.22	26.23	−104.08	−103.6	0.140	3.51
3	2	3	333.45	80.70	−332.03	−92.24	1.414	16.43
4	2	25	−246.11	89.31	250.55	−100.12	4.439	5.45
5	2	30	−250.00	−133.42	250.00	147.24	0.000	13.82
6	3	4	123.06	160.38	−122.51	−174.32	0.544	8.92
7	3	18	−73.03	−30.54	73.08	8.30	0.059	0.71
8	4	5	−196.54	−21.20	196.86	12.78	0.312	5.00
9	4	14	−220.94	−8.49	221.34	1.030	0.392	6.33
10	5	6	−521.95	−55.55	522.50	58.32	0.549	7.14
11	5	8	325.09	42.76	−324.23	−45.37	0.863	12.09
12	6	7	439.68	79.60	−438.49	−72.56	1.194	18.31
13	6	11	−279.50	−37.36	280.05	29.73	0.549	6.43
14	6	31	−682.68	−100.56	682.68	235.79	0.000	135.24
15	7	8	204.69	−11.44	−204.52	5.68	0.170	1.95
16	8	9	6.75	−136.92	−6.420	102.83	0.326	5.15
17	9	39	−0.08	−36.23	0.08	−91.73	0.007	0.19
18	10	11	309.18	59.26	−308.79	−62.53	0.389	4.18
19	10	13	340.82	74.67	−340.34	−76.98	0.477	5.13
20	10	32	−650.00	−133.93	650.00	232.33	0.000	98.4
21	12	11	−28.72	−31.98	28.75	32.80	0.030	0.81
22	12	13	−25.68	−30.39	25.71	31.08	0.026	0.70
23	13	14	314.63	45.89	−313.73	−53.12	0.904	10.14
24	14	15	−35.63	−89.16	35.75	53.33	0.113	1.36
25	15	16	−315.75	−176.33	316.86	169.98	1.114	11.63
26	16	17	177.78	−80.31	−177.54	68.94	0.242	3.07
27	16	19	−451.33	−47.15	454.39	51.36	3.059	37.28
28	16	21	−329.62	19.33	330.44	−32.75	0.820	13.84
29	16	24	−42.68	−94.16	42.71	87.39	0.028	0.55
30	17	18	191.32	6.86	−191.08	−18.30	0.238	2.79
31	17	27	26.22	−35.79	−26.20	1.090	0.012	0.16
32	19	20	174.72	−6.89	−174.5	11.18	0.218	4.29
33	19	33	−629.11	−44.47	632.00	103.03	2.886	58.55
34	20	34	−505.50	−114.18	508.00	164.25	2.503	50.07
35	21	22	−604.44	−82.25	607.21	102.85	2.770	48.47
36	22	23	42.79	41.60	−42.76	−61.50	0.025	0.39
37	22	35	−650.00	−144.44	650.00	204.75	0.000	60.31
38	23	24	353.84	−3.95	−351.31	4.81	2.522	40.13
39	23	36	−558.57	−19.16	560.00	96.81	1.428	77.66
40	25	26	63.79	−21.80	−63.68	−36.42	0.118	1.19
41	25	37	−538.34	74.72	540.00	−10.62	1.658	64.10
42	26	27	255.68	59.54	−254.8	−76.59	0.888	9.32
43	26	28	−140.79	−18.17	141.58	−59.77	0.790	8.70
44	26	29	−190.22	−21.95	192.14	−71.24	1.915	21.00
45	28	29	−347.58	32.17	349.13	−42.93	1.555	16.77
46	29	38	−824.77	87.27	830.00	14.76	5.232	102.02
						Total:	43.756	1009.53

Table A37. Results of the mixed AC/DC PF calculations in Case 7-2. DC bus data.

Bus DC	Bus Injection		Total Loss
	P	Q	P
1	−40.00	−20.00	1.24
2	−45.87	25.63	1.25
3	40.00	40.00	1.24
4	40.00	40.00	1.24
5	−128.03	−141.24	2.44
6	40.00	30.00	1.22
7	40.00	20.00	1.20
8	40.00	40.00	1.24
	Total:		11.07

Table A38. Results of the mixed AC/DC PF calculations in Case 7-2. DC branch data.

Branch	From	To	From Bus P	To Bus P	Power Losses P
1	1	2	−2.19	2.19	0.00
2	1	4	40.95	−40.51	0.44
3	2	3	42.43	−41.96	0.47
4	3	4	0.72	−0.72	0.00
5	5	6	41.86	−41.22	0.64
6	5	7	41.85	−41.21	0.64
7	5	8	41.87	−41.23	0.64
8	6	7	−0.01	0.01	0.00
9	6	8	0.01	−0.01	0.00
				Total:	2.83

Table A39. Droop control settings for each converter station.

Converter Station	Droop Parameter	P_{DC}^* (MW)	V_{DC}^* (p.u.)
1	0.0017	−50.0000	1.0079
2	0.0017	−35.0000	1.0000
3	0.0017	60.0000	1.0000
4	0.0017	60.0000	1.0000
5	0.0017	−35.0000	1.0000
6	0.0017	40.0000	1.0000
7	0.0017	40.0000	1.0000
8	0.0017	40.0000	1.0000

* P_{DC}^* and V_{DC}^* are the reference power and reference DC voltage, respectively.

Table A40. Results of Mixed AC/DC PF calculations in Case 7-3. AC bus data.

Bus	Voltage		Generation		Load	
	$|V|$ (p.u.)	θ (°)	P	Q	P	Q
1	1.047	−12.504	–	–	97.60	44.20
2	1.051	−9.194	–	–	–	–
3	1.034	−11.802	–	–	322.00	2.40
4	0.997	−12.917	–	–	500.00	184.00
5	1.001	−11.426	–	–	–	–
6	1.004	−10.647	–	–	–	–
7	0.994	−12.949	–	–	233.80	84.00
8	0.994	−13.501	–	–	522.00	176.60
9	1.037	−13.823	–	–	6.50	−66.60
10	1.012	−8.558	–	–	–	–
11	1.009	−9.310	–	–	–	–
12	1.000	−9.926	–	–	8.530	88.00
13	1.008	−9.343	–	–	–	–
14	1.000	−11.08	–	–	–	–
15	1.016	−11.186	–	–	320.00	153.00
16	1.034	−9.727	–	–	329.00	32.30
17	1.04	−10.668	–	–	–	–
18	1.037	−11.483	–	–	158.00	30.00
19	1.051	−5.110	–	–	–	–
20	0.991	−6.519	–	–	680.00	103.00
21	1.034	−7.328	–	–	274.00	115.00
22	1.051	−2.889	–	–	–	–
23	1.046	−3.087	–	–	247.50	84.60
24	1.040	−9.607	–	–	308.60	−92.20
25	1.060	−7.783	–	–	224.00	47.20
26	1.055	−8.920	–	–	139.00	17.00
27	1.042	−10.869	–	–	281.00	75.50
28	1.052	−5.418	–	–	206.00	27.60
29	1.051	−2.664	–	–	283.50	26.90
30	1.050	−6.785	250.00	146.82	–	–
31	0.982	0.000	689.99	240.44	9.200	4.60
32	0.984	−0.532	650.00	232.48	–	–
33	0.997	0.105	632.00	103.62	–	–
34	1.012	−1.329	508.00	164.52	–	–
35	1.049	2.067	650.00	205.41	–	–
36	1.064	4.758	560.00	97.19	–	–
37	1.028	−1.007	540.00	−9.75	–	–
38	1.027	4.395	830.00	15.39	–	–
39	1.030	−13.830	1000.00	54.04	1104.00	250.00
	Total:		6309.99	1250.16	6254.23	1387.10

Table A41. Results of Mixed AC/DC PF calculations in Case 7-3. AC branch data.

Branch	From	To	From Bus Injection		To Bus Injection		Power Losses	
			P	Q	P	Q	P	Q
1	1	2	−154.00	−31.28	154.76	−36.7	0.759	8.91
2	1	39	102.52	27.08	−102.38	−104.51	0.138	3.46
3	2	3	335.86	81.69	−334.43	−92.98	1.435	16.67
4	2	25	−240.62	88.03	244.87	−99.07	4.253	5.23
5	2	30	−250.00	−133.01	250.00	146.82	0.000	13.8
6	3	4	104.84	161.43	−104.34	−176.12	0.498	8.17
7	3	18	−46.55	−30.85	46.57	8.23	0.026	0.32
8	4	5	−203.98	−20.74	204.32	12.72	0.336	5.38
9	4	14	−247.94	−7.14	248.43	1.33	0.494	7.97
10	5	6	−529.57	−55.39	530.13	58.38	0.565	7.35
11	5	8	325.25	42.68	−324.39	−45.26	0.864	12.1
12	6	7	441.24	79.61	−440.04	−72.45	1.203	18.44
13	6	11	−290.58	−36.72	291.17	29.61	0.593	6.94
14	6	31	−680.79	−101.27	680.79	235.84	0.000	134.58
15	7	8	206.24	−11.55	−206.07	5.83	0.172	1.98
16	8	9	8.45	−137.17	−8.12	103.12	0.328	5.18
17	9	39	1.62	−36.52	−1.62	−91.44	0.007	0.18
18	10	11	317.57	59.28	−317.16	−62.33	0.409	4.4
19	10	13	332.43	74.78	−331.97	−77.32	0.455	4.9
20	10	32	−650.00	−134.06	650.00	232.48	0.000	98.41
21	12	11	−25.96	−31.97	25.99	32.72	0.027	0.75
22	12	13	−24.57	−30.42	24.59	31.09	0.025	0.67
23	13	14	307.38	46.23	−306.52	−53.90	0.864	9.7
24	14	15	2.66	−91.24	−2.56	55.21	0.096	1.16
25	15	16	−300.93	−178.21	301.97	171.1	1.040	10.86
26	16	17	192.66	−79.73	−192.39	68.82	0.277	3.52
27	16	19	−451.33	−47.94	454.39	52.19	3.061	37.3
28	16	21	−329.62	18.78	330.44	−32.19	0.820	13.84
29	16	24	−42.68	−94.52	42.71	87.76	0.028	0.56
30	17	18	188.28	6.71	−188.04	−18.23	0.231	2.7
31	17	27	20.60	−35.53	−20.59	0.80	0.009	0.12
32	19	20	174.72	−7.14	−174.50	11.44	0.218	4.29
33	19	33	−629.11	−45.05	632.00	103.62	2.887	58.57
34	20	34	−505.50	−114.44	508.00	164.52	2.504	50.08
35	21	22	−604.44	−82.81	607.21	103.44	2.771	48.5
36	22	23	42.79	41.63	−42.76	−61.53	0.025	0.39
37	22	35	−650.00	−145.07	650.00	205.41	0.000	60.34
38	23	24	353.84	−3.56	−351.31	4.44	2.523	40.14
39	23	36	−558.57	−19.51	560.00	97.19	1.428	77.67
40	25	26	69.47	−21.98	−69.33	−35.99	0.139	1.41
41	25	37	−538.34	73.85	540.00	−9.75	1.658	64.1
42	26	27	261.34	59.66	−260.41	−76.30	0.926	9.72
43	26	28	−140.79	−18.45	141.58	−59.46	0.789	8.70
44	26	29	−190.22	−22.23	192.13	−70.92	1.915	21.0
45	28	29	−347.58	31.86	349.13	−42.61	1.555	16.77
46	29	38	−824.77	86.63	830.00	15.39	5.232	102.03
						Total:	43.583	1009.26

Table A42. Results of the mixed AC/DC PF calculations in Case 7-3. DC bus data.

Bus DC	Bus Injection		Total Loss
	P	Q	P
1	−56.26	−20.00	1.30
2	−42.00	25.61	1.23
3	45.86	40.00	1.26
4	46.12	40.00	1.26
5	−55.43	−143.80	2.04
6	16.51	30.00	1.16
7	16.53	20.00	1.14
8	16.48	40.00	1.19
		Total:	10.58

Table A43. Results of the mixed AC/DC PF calculations in Case 7-3. DC branch data.

Branch			From Bus	To Bus	Power Losses
	From	To	P	P	P
1	1	2	5.27	−5.26	0.01
2	1	4	49.69	−49.03	0.65
3	2	3	46.03	−45.47	0.56
4	3	4	−1.65	1.65	0.00
5	5	6	17.79	−17.67	0.12
6	5	7	17.79	−17.67	0.12
7	5	8	17.79	−17.67	0.12
8	6	7	0.00	0.00	0.00
9	6	8	0.00	0.00	0.00
				Total:	1.58

Table A44. Results of Mixed AC/DC PF calculations in Case 7-4. AC bus data.

Bus	Voltage		Generation		Load	
	$|V|$ (p.u.)	θ (°)	P	Q	P	Q
1	1.043	−15.383	−	−	97.60	44.20
2	1.058	−11.984	−	−	−	−
3	1.055	−14.490	−	−	322.00	2.40
4	0.932	−16.184	−	−	500.00	184.00
5	0.963	−12.185	−	−	−	−
6	0.971	−10.906	−	−	−	−
7	0.961	−13.647	−	−	233.80	84.00
8	0.960	−14.378	−	−	522.00	176.60
9	1.023	−15.624	−	−	6.50	−66.6
10	0.995	−6.307	−	−	−	−
11	0.985	−7.818	−	−	−	−
12	0.975	−7.165	−	−	8.53	88.00
13	0.993	−6.380	−	−	−	−
14	0.994	−6.385	−	−	−	−
15	1.016	−14.260	−	−	320.00	153.00
16	1.033	−12.688	−	−	329.00	32.30
17	1.035	−13.530	−	−	−	−
18	1.032	−14.217	−	−	158.00	30.00
19	1.050	−8.066	−	−	−	−
20	0.991	−9.477	−	−	680.00	103.00
21	1.032	−10.284	−	−	274.00	115.00
22	1.050	−5.839	−	−	−	−
23	1.045	−6.037	−	−	247.50	84.60
24	1.038	−12.569	−	−	308.60	−92.20
25	1.064	−10.503	−	−	224.00	47.20
26	1.055	−11.705	−	−	139.00	17.00
27	1.040	−13.691	−	−	281.00	75.50
28	1.052	−8.202	−	−	206.00	27.60
29	1.051	−5.447	−	−	283.50	26.90
30	1.050	−9.591	250.00	108.31	−	−
31	0.982	0.000	683.41	362.78	9.20	4.60
32	0.984	1.860	650.00	313.16	−	−
33	0.997	−2.849	632.00	107.87	−	−
34	1.012	−4.287	508.00	166.49	−	−
35	1.049	−0.880	650.00	210.19	−	−
36	1.064	1.812	560.00	99.90	−	−
37	1.028	−3.746	540.00	−26.63	−	−
38	1.027	1.612	830.00	16.42	−	−
39	1.030	−16.197	1000.00	126.31	1104.00	250.00
Total:			6303.41	1484.80	6254.23	1387.10

Table A45. Results of Mixed AC/DC PF calculations in Case 7-4. AC branch data.

Branch	From	To	From Bus Injection		To Bus Injection		Power Losses	
			P	Q	P	Q	P	Q
1	1	2	−160.79	−56.77	161.63	−10.46	0.843	9.90
2	1	39	63.19	12.57	−63.13	−91.60	0.063	1.57
3	2	3	323.21	−12.20	−322.00	−2.40	1.213	14.09
4	2	25	−234.85	118.79	239.31	−129.74	4.458	5.48
5	2	30	−250.00	−96.12	250.00	108.31	0.000	12.19
6	3	4	0.00	0.00	0.00	0.00	0.000	0.00
7	3	18	0.00	0.00	0.00	0.00	0.000	0.00
8	4	5	−500.00	−184.00	502.60	213.48	2.595	41.53
9	4	14	0.00	0.00	0.00	0.00	0.000	0.00
10	5	6	−819.04	−214.86	820.59	230.88	1.544	20.08
11	5	8	316.45	1.38	−315.58	−2.93	0.864	12.10
12	6	7	490.24	81.71	−488.66	−68.05	1.579	24.20
13	6	11	−636.61	−105.52	639.69	128.36	3.084	36.12
14	6	31	−674.21	−207.08	674.21	358.18	0.000	151.11
15	7	8	254.86	−15.95	−254.58	12.01	0.282	3.25
16	8	9	48.16	−185.68	−47.40	160.29	0.763	12.05
17	9	39	40.90	−93.69	−40.87	−32.10	0.025	0.63
18	10	11	617.69	175.82	−616.01	−165.00	1.672	17.97
19	10	13	32.31	29.83	−32.31	−36.94	0.009	0.09
20	10	32	−650.00	−205.65	650.00	313.16	0.000	107.51
21	12	11	23.71	−35.79	−23.68	36.64	0.031	0.85
22	12	13	−32.24	−52.21	32.31	53.95	0.064	1.74
23	13	14	0.00	−17.01	0.00	0.00	0.001	0.01
24	14	15	0.00	0.00	0.00	0.00	0.000	0.00
25	15	16	−320.00	−153.00	321.07	146.27	1.074	11.21
26	16	17	173.51	−42.68	−173.31	30.96	0.206	2.62
27	16	19	−451.30	−53.64	454.38	58.17	3.077	37.50
28	16	21	−329.60	14.83	330.42	−28.14	0.821	13.85
29	16	24	−42.68	−97.08	42.71	90.37	0.030	0.58
30	17	18	158.17	17.88	−158.00	−30.00	0.168	1.96
31	17	27	15.14	−48.84	−15.12	14.44	0.015	0.20
32	19	20	174.73	−8.99	−174.51	13.30	0.218	4.30
33	19	33	−629.11	−49.17	632.00	107.87	2.894	58.70
34	20	34	−505.49	−116.30	508.00	166.49	2.510	50.20
35	21	22	−604.42	−86.86	607.21	107.73	2.782	48.69
36	22	23	42.79	41.86	−42.77	−61.73	0.025	0.40
37	22	35	−650.00	−149.59	650.00	210.19	0.000	60.60
38	23	24	353.84	−0.77	−351.31	1.83	2.529	40.23
39	23	36	−558.57	−22.10	560.00	99.90	1.430	77.80
40	25	26	75.03	−8.33	−74.86	−49.50	0.173	1.74
41	25	37	−538.34	90.86	540.00	−26.63	1.661	64.23
42	26	27	266.87	74.07	−265.88	−89.94	0.992	10.42
43	26	28	−140.79	−18.90	141.58	−58.95	0.789	8.70
44	26	29	−190.21	−22.67	192.13	−70.41	1.915	20.99
45	28	29	−347.58	31.35	349.14	−42.10	1.555	16.78
46	29	38	−824.77	85.61	830.00	16.42	5.232	102.03
						Total:	49.186	1106.20

Table A46. Results of Mixed AC/DC PF calculations in Case 7-5. AC bus data.

Bus	Voltage		Generation		Load	
	$|V|$ (p.u.)	θ (°)	P	Q	P	Q
1	1.046	−5.461	–	–	97.60	44.20
2	1.060	1.265	–	–	–	–
3	1.063	−0.937	–	–	322.00	2.40
4	0.934	−16.570	–	–	500.00	184.00
5	0.969	−12.289	–	–	–	–
6	0.977	−11.124	–	–	–	–
7	0.967	−13.306	–	–	233.80	84.00
8	0.966	−13.764	–	–	522.00	176.60
9	1.025	−10.911	–	–	6.50	−66.60
10	1.003	−7.787	–	–	–	–
11	0.994	−8.935	–	–	–	–
12	1.000	−9.246	–	–	8.53	88.00
13	1.002	−8.201	–	–	–	–
14	1.000	−8.935	–	–	–	–
15	1.032	4.023	–	–	320.00	153.00
16	1.045	5.368	–	–	329.00	32.30
17	1.051	4.342	–	–	–	–
18	1.050	3.841	–	–	158.00	30.00
19	1.055	9.953	–	–	–	–
20	0.993	8.557	–	–	680.00	103.00
21	1.041	7.738	–	–	274.00	115.00
22	1.055	12.140	–	–	–	–
23	1.050	11.944	–	–	247.50	84.60
24	1.049	5.488	–	–	308.60	−92.20
25	1.070	3.371	–	–	224.00	47.20
26	1.064	4.157	–	–	139.00	17.00
27	1.053	3.112	–	–	281.00	75.50
28	1.057	7.633	–	–	206.00	27.60
29	1.054	10.373	–	–	283.50	26.90
30	1.050	3.655	250.00	99.19	–	–
31	0.982	0.000	701.04	343.28	9.20	4.60
32	0.984	0.317	650.00	277.42	–	–
33	0.997	15.160	632.00	78.96	–	–
34	1.012	13.742	508.00	153.10	–	–
35	1.049	17.078	650.00	177.74	–	–
36	1.064	19.762	560.00	81.49	–	–
37	1.028	10.095	540.00	−55.46	–	–
38	1.026	17.419	830.00	−5.42	–	–
39	1.030	−8.849	1000.00	115.54	1104.00	250.00
Total:			6321.04	1265.84	6254.23	1387.10

Table A47. Results of Mixed AC/DC PF calculations in Case 7-5. AC branch data.

Branch	From	To	From Bus Injection		To Bus Injection		Power Losses	
			P	Q	P	Q	P	Q
1	1	2	−314.89	−27.33	318.06	−12.82	3.176	37.29
2	1	39	257.29	23.13	−256.65	−87.88	0.643	16.07
3	2	3	282.95	−55.56	−282.00	37.60	0.947	11.00
4	2	25	−351.01	155.69	360.37	−160.75	9.357	11.50
5	2	30	−250.00	−87.31	250.00	99.19	0.000	11.88
6	3	4	0.00	0.00	0.00	0.00	0.000	0.00
7	3	18	0.00	0.00	0.00	0.00	0.000	0.00
8	4	5	−540.00	−204.00	543.03	240.39	3.034	48.54
9	4	14	0.00	0.00	0.00	0.00	0.000	0.000
10	5	6	−758.72	−243.95	760.07	257.42	1.352	17.58
11	5	8	215.68	3.56	−215.29	−11.80	0.398	5.57
12	6	7	396.78	85.56	−395.74	−80.25	1.042	15.98
13	6	11	−465.01	−158.13	466.76	165.20	1.755	20.55
14	6	31	−691.84	−184.85	691.84	338.68	0.000	153.82
15	7	8	161.94	−3.75	−161.82	−2.25	0.112	1.29
16	8	9	−144.89	−162.55	145.92	141.15	1.034	16.33
17	9	39	−152.42	−74.55	152.65	−46.59	0.222	5.56
18	10	11	480.08	167.66	−479.05	−163.81	1.034	11.11
19	10	13	169.92	6.61	−169.8	−12.69	0.115	1.24
20	10	32	−650.00	−174.27	650.00	277.42	0.000	103.15
21	12	11	−12.28	1.46	12.29	−1.39	0.002	0.07
22	12	13	−42.29	−15.31	42.32	16.20	0.033	0.89
23	13	14	127.48	−3.51	−127.33	−12.11	0.146	1.64
24	14	15	0.00	0.00	0.00	0.00	0.000	0.00
25	15	16	−280.00	−123.00	280.77	112.64	0.773	8.07
26	16	17	214.10	−92.92	−213.76	82.52	0.341	4.34
27	16	19	−451.47	−14.50	454.46	17.42	2.988	36.42
28	16	21	−329.71	41.95	330.53	−55.83	0.820	13.83
29	16	24	−42.70	−79.48	42.72	72.44	0.021	0.41
30	17	18	118.09	−3.51	−118.00	−10.00	0.089	1.04
31	17	27	135.67	−39.01	−135.45	6.39	0.222	2.96
32	19	20	174.69	3.61	−174.47	0.65	0.216	4.26
33	19	33	−629.14	−21.03	632.00	78.96	2.856	57.93
34	20	34	−505.53	−103.65	508.00	153.10	2.472	49.45
35	21	22	−604.53	−59.17	607.24	78.50	2.714	47.49
36	22	23	42.76	40.27	−42.73	−60.33	0.024	0.38
37	22	35	−650.00	−118.77	650.00	177.74	0.000	58.97
38	23	24	353.82	−19.77	−351.32	19.76	2.499	39.75
39	23	36	−558.58	−4.49	560.00	81.49	1.415	77.00
40	25	26	−46.04	−6.66	46.12	−53.07	0.075	0.76
41	25	37	−538.33	120.21	540.00	−55.46	1.675	64.75
42	26	27	145.88	58.48	−145.55	−81.89	0.327	3.43
43	26	28	−140.68	−9.26	141.48	−69.69	0.797	8.79
44	26	29	−190.31	−13.14	192.24	−81.25	1.925	21.10
45	28	29	−347.48	42.09	349.03	−53.07	1.554	16.76
46	29	38	−824.77	107.42	830.00	−5.42	5.231	102.00
						Total:	53.436	1110.95

Table A48. Results of the mixed AC/DC PF calculations in Case 7-5. DC bus data.

Bus DC	Bus Injection		Total Loss
	P	Q	P
1	−40.00	−20.00	1.25
2	−46.04	74.15	1.42
3	40.00	40.00	1.23
4	40.00	40.00	1.24
5	−127.33	−12.11	1.75
6	40.00	30.00	1.22
7	40.00	20.00	1.2
8	40.00	40.00	1.24
		Total:	10.55

Table A49. Results of the mixed AC/DC PF calculations in Case 7-5. DC branch data.

Branch			From Bus	To Bus	Power Losses
	From	To	P	P	P
1	1	2	−2.20	2.20	0.00
2	1	4	40.95	−40.51	0.44
3	2	3	42.43	−41.96	0.47
4	3	4	0.73	−0.73	0.00
5	5	6	41.86	−41.22	0.64
6	5	7	41.85	−41.21	0.64
7	5	8	41.87	−41.23	0.64
8	6	7	−0.01	0.01	0.00
9	6	8	0.01	−0.01	0.00
				Total:	2.83

Table A50. Results of Mixed AC/DC PF calculations in Case 7-6. AC bus data.

Bus	Voltage		Generation		Load	
	$\lvert V\rvert$ (p.u.)	θ (°)	P	Q	P	Q
1	1.049	−8.267	−	−	97.60	44.20
2	1.062	−2.889	−	−	−	−
3	1.066	−5.034	−	−	322.00	2.40
4	0.932	−16.723	−	−	500.00	184.00
5	0.967	−12.294	−	−	−	−
6	0.975	−11.059	−	−	−	−
7	0.965	−13.409	−	−	233.80	84.00
8	0.965	−13.950	−	−	522.00	176.60
9	1.025	−12.240	−	−	6.50	−66.60
10	1.002	−7.176	−	−	−	−
11	0.993	−8.494	−	−	−	−
12	1.000	−8.589	−	−	8.53	88.00
13	1.001	−7.424	−	−	−	−
14	1.000	−7.736	−	−	−	−
15	1.03	−3.037	−	−	320.00	153.00
16	1.044	−1.572	−	−	329.00	32.30
17	1.050	−2.488	−	−	−	−
18	1.048	−3.091	−	−	158.00	30.00
19	1.054	3.016	−	−	−	−
20	0.993	1.618	−	−	680.00	103.00
21	1.040	0.800	−	−	274.00	115.00
22	1.054	5.205	−	−	−	−
23	1.05	5.008	−	−	247.50	84.60
24	1.048	−1.452	−	−	308.60	−92.20
25	1.069	−1.128	−	−	224.00	47.20
26	1.063	−1.501	−	−	139.00	17.00
27	1.052	−3.084	−	−	281.00	75.50
28	1.056	1.978	−	−	206.00	27.60
29	1.054	4.719	−	−	283.50	26.90
30	1.050	−0.506	250.00	82.91	−	−
31	0.982	0.000	696.02	347.70	9.20	4.60
32	0.984	0.934	650.00	280.53	−	−
33	0.997	8.223	632.00	80.97	−	−
34	1.012	6.803	508.00	154.03	−	−
35	1.049	10.145	650.00	180.00	−	−
36	1.064	12.829	560.00	82.78	−	−
37	1.028	5.600	540.00	−52.18	−	−
38	1.027	11.767	830.00	−3.06	−	−
39	1.030	−10.916	1000.00	99.46	1104.00	250.00
	Total:		6316.02	1253.14	6254.23	1387.10

Table A51. Results of Mixed AC/DC PF calculations in Case 7-6. AC branch data.

Branch	From	To	From Bus Injection		To Bus Injection		Power Losses	
			P	Q	P	Q	P	Q
1	1	2	−254.22	−39.12	256.28	−14.61	2.056	24.14
2	1	39	202.75	34.92	−202.32	−105.31	0.426	10.66
3	2	3	277.03	−56.22	−276.13	37.60	0.904	10.50
4	2	25	−283.31	142.35	289.70	−151.10	6.384	7.84
5	2	30	−250.00	−71.52	250.00	82.91	0.000	11.39
6	3	4	0.00	0.00	0.00	0.00	0.000	0.00
7	3	18	0.00	0.00	0.00	0.00	0.000	0.00
8	4	5	−556.28	−204.00	559.49	243.27	3.211	51.37
9	4	14	0.00	0.00	0.00	0.00	0.000	0.00
10	5	6	−800.66	−243.89	802.15	259.24	1.496	19.45
11	5	8	241.17	0.62	−240.67	−7.41	0.498	6.97
12	6	7	425.54	84.41	−424.34	−76.76	1.193	18.29
13	6	11	−540.87	−153.36	543.18	166.98	2.311	27.07
14	6	31	−686.82	−190.29	686.82	343.10	0.000	152.81
15	7	8	190.54	−7.24	−190.39	1.78	0.156	1.79
16	8	9	−90.94	−170.96	91.73	145.68	0.785	12.39
17	9	39	−98.23	−79.08	98.32	−45.23	0.094	2.36
18	10	11	548.16	170.50	−546.84	−163.59	1.318	14.17
19	10	13	101.84	6.52	−101.80	−13.39	0.042	0.45
20	10	32	−650.00	−177.03	650.00	280.53	0.000	103.50
21	12	11	−3.66	3.41	3.66	−3.40	0.000	0.01
22	12	13	−47.06	−14.10	47.10	15.16	0.039	1.06
23	13	14	54.70	−1.77	−54.67	−15.17	0.027	0.31
24	14	15	0.00	0.00	0.00	0.00	0.000	0.00
25	15	16	−303.49	−123.00	304.38	113.91	0.891	9.30
26	16	17	190.48	−88.27	−190.20	77.07	0.275	3.50
27	16	19	−451.46	−17.25	454.45	20.27	2.993	36.48
28	16	21	−329.70	40.04	330.52	−53.88	0.819	13.83
29	16	24	−42.70	−80.72	42.72	73.70	0.021	0.42
30	17	18	141.60	−3.02	−141.47	−10.00	0.128	1.49
31	17	27	65.09	−34.05	−65.04	−0.74	0.053	0.71
32	19	20	174.69	2.73	−174.47	1.53	0.216	4.26
33	19	33	−629.14	−23.00	632.00	80.97	2.858	57.97
34	20	34	−505.53	−104.53	508.00	154.03	2.475	49.50
35	21	22	−604.52	−61.12	607.24	80.55	2.718	47.57
36	22	23	42.76	40.38	−42.74	−60.43	0.024	0.38
37	22	35	−650.00	−120.93	650.00	180.00	0.000	59.07
38	23	24	353.82	−18.44	−351.32	18.50	2.500	39.78
39	23	36	−558.58	−5.73	560.00	82.78	1.416	77.05
40	25	26	24.63	−12.96	−24.61	−47.17	0.025	0.26
41	25	37	−538.33	116.86	540.00	−52.18	1.673	64.68
42	26	27	216.60	54.67	−215.96	−74.76	0.638	6.70
43	26	28	−140.69	−10.32	141.49	−68.53	0.796	8.78
44	26	29	−190.30	−14.18	192.23	−80.07	1.923	21.09
45	28	29	−347.49	40.93	349.04	−51.88	1.554	16.76
46	29	38	−824.77	105.05	830.00	−3.06	5.230	101.99
						Total:	50.166	1098.10

Table A52. Results of the mixed AC/DC PF calculations in Case 7-6. DC bus data.

Bus DC	Bus Injection		Total Loss
	P	Q	P
1	−56.28	−20.00	1.32
2	−42.19	77.31	1.42
3	45.87	40.00	1.25
4	46.12	40.00	1.26
5	−54.67	−15.17	1.29
6	16.51	30.00	1.16
7	16.53	20.00	1.14
8	16.49	40.00	1.19
Total:			10.03

Table A53. Results of the mixed AC/DC PF calculations in Case 7-6. DC branch data.

Branch			From Bus	To Bus	Power Losses
	From	To	P	P	P
1	1	2	5.27	−5.26	0.01
2	1	4	49.69	−49.03	0.65
3	2	3	46.03	−45.47	0.56
4	3	4	−1.65	1.65	0.00
5	5	6	17.79	−17.67	0.12
6	5	7	17.79	−17.67	0.12
7	5	8	17.79	−17.67	0.12
8	6	7	0.00	0.00	0.00
9	6	8	0.00	0.00	0.00
				Total:	1.58

References

1. Mohammadi, F.; Nazri, G.-A.; Saif, M. A Bidirectional Power Charging Control Strategy for Plug-in Hybrid Electric Vehicles. *Sustainability* **2019**, *11*, 4317. [CrossRef]
2. Bahrman, M.P.; Johnson, B.K. The ABCs of HVDC Transmission Technologies. *IEEE Power Energy Mag.* **2007**, *5*, 32–44. [CrossRef]
3. Flourentzou, N.; Agelidis, V.G.; Demetriades, G.D. VSC-Based HVDC Power Transmission Systems: An Overview. *IEEE Trans. Power Electron.* **2009**, *24*, 592–602. [CrossRef]
4. Mohammadi, F.; Nazri, G.-A.; Saif, M. An Improved Droop-Based Control Strategy for MT-HVDC Systems. *Electronics* **2020**, *9*, 87.
5. Xu, L.; Fan, L. Impedance-Based Resonance Analysis in a VSC-HVDC System. *IEEE Trans. Power Deliv.* **2013**, *28*, 2209–2216.
6. Xu, L.; Fan, L.; Miao, Z. DC Impedance-Model-Based Resonance Analysis of a VSC-HVDC System. *IEEE Trans. Power Deliv.* **2015**, *30*, 1221–1230. [CrossRef]
7. Pandya, K.S.; Joshi, S.K. A Survey of Optimal Power Flow Methods. *J. Theor. Appl. Inf. Technol.* **2008**, *4*, 450–458.
8. Frank, S.; Steponavice, I.; Rebennack, S. Optimal Power Flow: A Bibliographic Survey II: Non-Deterministic and Hybrid Methods. *Energy Syst.* **2012**, *3*, 259–289. [CrossRef]
9. Mohammadi, F.; Nazri, G.-A.; Saif, M. A New Topology of a Fast Proactive Hybrid DC Circuit Breaker for MT-HVDC Grids. *Sustainability* **2019**, *11*, 4493. [CrossRef]
10. Mohammadi, F.; Zheng, C. Stability Analysis of Electric Power System. In Proceedings of the 4th National Conference on Technology in Electrical and Computer Engineering, Tehran, Iran, 27 December 2018.
11. Mohammadi, F.; Nazri, G.-A.; Saif, M. A Fast Fault Detection and Identification Approach in Power Distribution Systems. In Proceedings of the 5th International Conference on Power Generation Systems and Renewable Energy Technologies (PGSRET), Istanbul, Turkey, 26–27 August 2019.

12. Arrillaga, J.; Bodger, P. Integration of HVDC Links with Fast-Decoupled Load-Flow Solutions. *Proc. Inst. Electr. Eng.* **1977**, *124*, 463–468. [CrossRef]

13. Baradar, M.; Ghandhari, M.; van Hertem, D. The Modeling Multi-Terminal VSC-HVDC in Power Flow Calculation using Unified Methodology. In Proceedings of the 2nd IEEE PES International Conference and Exhibition on Innovative Smart Grid Technologies, Manchester, UK, 5–7 December 2011.

14. Baradar, M.; Ghandhari, M. A Multi-Option Unified Power-Flow Approach for Hybrid AC/DC Grids Incorporating Multi-terminal VSC-HVDC. *IEEE Trans. Power Syst.* **2013**, *28*, 2376–2383. [CrossRef]

15. El-Hawary, M.E.; Ibrahim, S.T. A New Approach to AC-DC Load Flow Analysis. *Electr. Power Syst. Res.* **1995**, *33*, 193–200. [CrossRef]

16. Beerten, J.; van Hertem, D.; Belmans, R. VSC-MTDC Systems with a Distributed DC Voltage Control-A Power Flow Approach. In Proceedings of the 2011 IEEE Trondheim PowerTech, Trondheim, Norway, 19–23 June 2011.

17. Haileselassie, T.M.; Uhlen, K. Power Flow Analysis of Multi-Terminal HVDC Networks. In Proceedings of the 2011 IEEE Trondheim PowerTech, Trondheim, Norway, 19–23 June 2011.

18. Beerten, J.; Cole, S.; Belmans, R. Implementation Aspects of a Sequential AC/DC Power Flow Computation Algorithm for Multi-Terminal VSC-HVDC Systems. In Proceedings of the 9th IET International Conference on AC and DC Power Transmission, London, UK, 19–21 October 2010.

19. Beerten, J.; Cole, S.; Belmans, R. A Sequential AC/DC Power Flow Algorithm for Networks Containing Multi-Terminal VSC-HVDC Systems. In Proceedings of the IEEE PES General Meeting, Providence, RI, USA, 25–29 July 2010.

20. Beerten, J.; Cole, S.; Belmans, R. Generalized Steady-State VSC MTDC Model for Sequential AC/DC Power Flow Algorithms. *IEEE Trans. Power Syst.* **2012**, *27*, 821–829. [CrossRef]

21. Gonzalez-Longatt, F.; Roldan, J.; Charalambous, C.A. Power Flow Solution on Multi-Terminal HVDC Systems: Supergrid Case. In Proceedings of the International Conference on Renewable Energies and Power Quality, Santiago de Compostela, Galicia, Spain, 28–30 March 2012.

22. Hendriks, R.L.; Paap, G.C.; Kling, W.L. Control of a Multi-Terminal VSC Transmission Scheme for Interconnecting Offshore Wind Farms. In Proceedings of the European Wind Energy Conference and Exhibition, Milan, Italy, 7–10 May 2007.

23. Yang, Z.; Zhong, H.; Bose, A.; Xia, Q.; Kang, C. Optimal Power Flow in AC-DC Grids with Discrete Control Devices. *IEEE Trans. Power Syst.* **2018**, *33*, 1461–1472.

 *applied sciences*

Article

Multi-Objective Optimal Reactive Power Planning under Load Demand and Wind Power Generation Uncertainties Using ε-Constraint Method

Amir Hossein Shojaei [1], Ali Asghar Ghadimi [2], Mohammad Reza Miveh [3], Fazel Mohammadi [4,*] and Francisco Jurado [5]

[1] Department of Electrical and Computer Engineering, Payam-e-Golpayegan Higher Education Institute, Golpayegan, Iran; amirehosseinshojaie@yahoo.com
[2] Department of Electrical Engineering, Faculty of Engineering, Arak University, Arak 38156-8-8349, Iran; a-ghadimi@araku.ac.ir
[3] Department of Electrical Engineering, Tafresh University, Tafresh 39518-79611, Iran; miveh@tafreshu.ac.ir
[4] Electrical and Computer Engineering (ECE) Department, University of Windsor, Windsor, ON N9B 1K3, Canada
[5] Department of Electrical Engineering, University of Jaen, 23700 Linares, Spain; fjurado@ujaen.es
* Correspondence: fazel@uwindsor.ca or fazel.mohammadi@ieee.org

Received: 14 March 2020; Accepted: 16 April 2020; Published: 20 April 2020

Abstract: This paper presents an improved multi-objective probabilistic Reactive Power Planning (RPP) in power systems considering uncertainties of load demand and wind power generation. The proposed method is capable of simultaneously (1) reducing the reactive power investment cost, (2) minimizing the total active power losses, (3) improving the voltage stability, and (4) enhancing the loadability factor. The generators' voltage magnitude, the transformer's tap settings, and the output reactive power of VAR sources are taken into account as the control variables. To solve the probabilistic multi-objective RPP problem, the ε-constraint method is used. To test the effectiveness of the proposed approach, the IEEE 30-bus test system is implemented in the GAMS environment under five different conditions. Finally, for a better comprehension of the obtained results, a brief comparison of outcomes is presented.

Keywords: ε-Constraint method; multi-objective optimization; reactive power planning (RPP); uncertainty; wind farms

---

1. Introduction

Reactive Power Planning (RPP) in power systems can be considered as one of the most difficult and complicated problems due to its complex variables, constraints, and optimization algorithms [1]. It is related to optimal sizing and allocation of VAR sources in power systems to satisfy prescheduled objectives, such as determining the optimal allocation and minimizing the operation costs [2,3]. The main objective of RPP is to achieve feasible operation with a satisfactory voltage profile with a lack of VAR support conditions. According to the concept of VAR planning in power systems, various objectives functions can be defined for the RPP problem. These objectives may consist of cost-based objective functions or objective functions that maximize or minimize indices, such as voltage stability margin or system loadability [4,5]. Moreover, it is possible to express the RPP as a multi-objective optimization problem, which optimizes several goals simultaneously [1].

Moreover, there is an increasing interest in using Renewable Energy Resources (RESs), such as wind farms and solar power plants, in power systems due to their technical, environmental, and economic advantages [4–6]. However, with the high penetration of RESs in power systems, the challenges associated with RPP are dramatically increased. One of the main challenges that can affect the RPP is the uncertainty in the generation availability of RESs. Uncertainty in the sources' parameters leads to difficulties in proper decision-making in the planning of power systems. Furthermore, owing to the stochastic nature of the load demands in electric power systems, additional uncertainties should be considered in RPP.

In RPP research studies, the probabilistic decision-making process based on either source uncertainty or load demand uncertainty is a well-developed research topic. Nevertheless, the probabilistic multi-objective RPP in power systems considering the uncertainty of loads and wind farms at the same time has not been fully investigated. In [7], a novel approach for dynamic VAR planning to improve the short-term voltage stability and transient stability is proposed. The impact of FACTS devices in RPP is analyzed in [8,9]. However, in both studies, an attempt has been made to explain the problem in a deterministic context. A multi-objective RPP that mainly focuses on voltage stability is introduced in [3]. Nonetheless, it is modeled based on a deterministic approach. In [10], a multi-objective approach for RPP with wind generations is presented. In this study, various objectives, such as system loadability, power losses, and cost of reactive power investment are considered. In [11], the RPP is solved using the Genetic Algorithm (GA) to reach coordination in controlling the reactive power in the presence of wind farms and FACTS devices. The loadability factor of the system is optimized by the optimal allocation of wind farms and FACTS devices. This procedure is implemented when loads with constant Power Factor (PF) and wind farms without uncertainty are assumed. Using the Benders decomposition method and considering the high penetration of wind generation, the RPP problem is tackled as two-stage stochastic programming in [12]. Using the Differential Evolutionary Algorithm (DEA), the RPP is solved in a wind integrated system in [13]. A major problem with the suggested model is that it only includes the uncertainty in wind power generation. In [14], a multi-stage stochastic model for RPP is extended, which involves the uncertainty of loads. Nonetheless, the proposed model describes the probabilistic behavior of the system in the absence of wind farms. In [15], a mixed-integer quadratic model for long term VAR planning is proposed. An attempt was made to minimize the operation and investment cost of new VAR sources and the load shedding risk through multi-objective optimization. Though the uncertainty in demand is completely taken into account, the proposed model does not consider uncertainty in wind power generation. In [16], a stochastic model based on chanced constrained programming for RPP is defined. The proposed model is solved using GA. Although the uncertainty is modeled in the power generation, it optimizes only one objective, including operational and investment costs. A chanced constrained model is proposed for probabilistic RPP in [17]. The proposed model is solved through two-stage stochastic programming. The main disadvantage of the proposed model is that it only considers the load as a random parameter. Besides, only the investment cost of new VAR sources is taken into account as the main objective function.

The main drawback of all the mentioned research studies is that the optimal RPP considering load demand and wind power generation uncertainties at the same time are not fully investigated. This paper aims to address RPP as a probabilistic multi-objective problem in order to reduce the total cost of reactive power investment, minimize the active power losses, maximize the voltage stability index, and improve the loadability factor. The generators' voltage magnitude, the transformers tap settings, and the output reactive power of the VAR sources are considered as the main control variables. To cope with the probabilistic multi-objective RPP problem, the ε-constraint technique is employed. To validate the efficiency of the proposed method, the IEEE 30-bus test system is implemented in the GAMS environment under five various conditions. The obtained results show the effectiveness and high accuracy of the proposed method. Table 1 shows a comparison between the proposed probabilistic multi-objective RPP and the previously published research papers.

The rest of this paper is organized as follows. Section 2 deals with the uncertainty modeling. Problem formulation is presented in Section 3. Section 4 describes the optimization method. Simulation results are given in Section 5. Finally, some brief conclusions are summarized in Section 6.

Table 1. Comparison between the previous research studies and the proposed method.

Reference	Problem Framework	Load Demand Uncertainty	Wind Power Generation Uncertainty	Load Demand and Wind Power Generation Uncertainties	Objective Function	Solution Methodology
[3]	Deterministic, Multi-Objective	-	-	-	Active power losses, Total VAR cost, Voltage stability index	MOEA
[4]	Deterministic, Multi-Objective	-	-	-	Investment cost, Short-term voltage stability level, Transient stability level	MOEA
[5]	Deterministic, Single-Objective	-	-	-	Active power losses, Voltage deviations, Operating cost	WOA, DE, GWO, QODE, QOGWO
[6]	Deterministic, Single-Objective	-	-	-	Active power losses, Operating cost	SPSO, APSO, EPSO
[7]	Deterministic, Multi-Objective	-	-	-	Investment cost, Short-term voltage stability, Transient stability	MOEA
[8]	Deterministic, Single-objective	-	-	-	Active power losses, Voltage deviations, Operating cost	WOA, DE, GWO, QODE, QOGWO
[9]	Deterministic, Single-Objective	-	-	-	Active power losses, Operating cost	SPSO, APSO, EPSO
[10]	Deterministic, Multi-Objective	-	-	-	Loadability factor, Active power losses, VAR investment cost	GA
[11]	Deterministic, Single-Objective	-	-	-	Loadability factor	GA
[12]	Probabilistic, Single-Objective	✔	✔	✔	Fuel cost, VAR cost, Total cost	Mathematical Programming
[13]	Probabilistic, Single-Objective	-	✔	-	VAR investment cost	DE
[14]	Probabilistic, Single-Objective	✔	-	-	VAR investment cost	Mathematical Programming
[15]	Probabilistic, Multi-Objective	✔	-	-	Operating cost, VAR investment cost, Load shedding risk	Multi-Objective Mathematical Programming (ε-constraint method)
[16]	Probabilistic, Single-Objective	✔	-	-	Operating cost, VAR investment cost	GA
[17]	Probabilistic, Single- Objective	✔	-	-	VAR investment cost	Mathematical Programming
Present Paper	Probabilistic, Multi-Objective	✔	✔	✔	Active power losses, Total VAR cost, Voltage stability index, Loadability factor	Multi-objective Mathematical Programming (ε-constraint method)

2. Uncertainty Modeling

In this section, the uncertainties in load demand and wind power generation in the RPP problem are modeled to cope with the stochastic nature of the load demand and wind power generation. In the following subsections, modeling of both the load demand and wind power uncertainties are described. Finally, modeling of the system uncertainty via scenario generation is presented.

2.1. Modeling the Load Demand Uncertainty

The uncertainty of the load is usually modeled by the normal distribution with mean (μ) and standard deviation (σ) [18]. In this paper, it is assumed that all the loads have constant PF, the same mean, and standard deviation. Therefore, for simplicity, a normal distribution is applied at the load level (λ) instead of applying in each load independently. The probability of each load level is shown by (π_l), and is calculated using Equation (1). The associated value of each load level is denoted by (λ_l), and can be obtained using Equation (2) [19]. It is worth mentioning that $\lambda_{Min,l}$ and $\lambda_{Max,l}$ are known as the minimum and maximum levels of the system loading at the lth load level, respectively.

$$\pi_l = \int_{\lambda_{Min,l}}^{\lambda_{Max,l}} \frac{1}{\sqrt{2\pi\sigma^2}} \exp\left(-\frac{(\lambda-\mu)^2}{2\sigma^2}\right) d\lambda \tag{1}$$

$$\lambda_l = \frac{1}{\pi_l} \int_{\lambda_{Min,l}}^{\lambda_{Max,l}} \lambda \frac{1}{\sqrt{2\pi\sigma^2}} \exp\left(-\frac{(\lambda-\mu)^2}{2\sigma^2}\right) d\lambda \tag{2}$$

2.2. Modeling the Wind Power Generation Uncertainty

Considering the intermittent nature of the wind speed, the Weibull distribution is often considered as the probability density function that can approximate the behavior of the wind with a reasonable error. Therefore, by defining the Weibull distribution for wind speed, the probability of wind speed at different intervals (scenarios) can easily be calculated. Equation (3) is a general expression for Weibull distribution [20]. Equation (4) can be used to calculate the probability of a wind speed interval (scenario). The corresponding value of each wind speed interval can be achieved using Equation (5).

$$PDF(v) = \frac{\beta}{\alpha}\left(\frac{v}{\alpha}\right)^{\beta-1} \exp\left(-\left(\frac{v}{\alpha}\right)^{\beta}\right) \tag{3}$$

$$\pi_w = \int_{v_{i,w}}^{v_{f,w}} \frac{\beta}{\alpha}\left(\frac{v}{\alpha}\right)^{\beta-1} \exp\left(-\left(\frac{v}{\alpha}\right)^{\beta}\right) dv \tag{4}$$

$$v_w = \frac{1}{\pi_w} \int_{v_{i,w}}^{v_{f,w}} v\frac{\beta}{\alpha}\left(\frac{v}{\alpha}\right)^{\beta-1} \exp\left(-\left(\frac{v}{\alpha}\right)^{\beta}\right) dv \tag{5}$$

where v denotes the wind speed, and α and β are the wind speed parameters that vary depending on the region in which the wind blows. Considering $v_{i,w}$ and $v_{f,w}$ as the initial speed and final speed of the hypothetical scenarios for wind speed, the probability (π_w) of occurrence of any wind speed scenario can simply be obtained. Thereafter, the wind speed (v_w) associated with each scenario is gained using the calculated probabilities.

The output power of a wind turbine is highly dependent on wind speed. Therefore, any wind turbine has a characteristic named power curve that exactly shows the capability of a wind turbine in power generation versus existing wind speed. Knowing a specific wind speed (v_w), one can estimate the output power of a wind turbine (P_w^{est}) through its power curve. The power curve is generally defined by a set of equations as it is stated in Equation (6) [21], which in terms, (v_{in}^c), (v_{rated}), and (v_{out}^c) denote the cut-in wind speed, rated wind speed, and cut-out wind speed for a wind turbine, respectively. The rated power (P_w^r) and the estimated output power of the wind turbine are also evident from Equation (6).

$$P_w^{est} = \begin{cases} 0, & v_w \leq v_{in}^c \\ \frac{v_w - v_{in}^c}{v_{rated} - v_{in}^c}, & v_{in}^c < v_w < v_{rated} \\ P_w^r, & v_{rated} < v_w < v_{out}^c \\ 0, & v_w \geq v_{out}^c \end{cases} \qquad (6)$$

In most research studies, the concept of the power curve is extended to a wind farm. Hence, instead of studying a single wind turbine, it is preferable to focus on a group of wind turbines that are in a special area and usually known as wind farms.

Considering several scenarios for a probabilistic problem is generally not an easy procedure. Depending on the problem type, various methods exist for scenario generation [22,23]. However, in this paper, a technique based on [19,24,25] is applied to generate a desirable number of scenarios with reasonable accuracy. In order to have a combination of load and wind scenarios, the following steps are taken:

1. Several scenarios for the load level are considered.
2. The probability of each system loading scenario (level of the load) and its corresponding value using Equations (1) and (2) are calculated.
3. Several scenarios for wind speed are considered.
4. The probability of each wind speed scenario and its corresponding value using Equations (4) and (5) are calculated.
5. The output power of the wind farm using the estimated wind speed in each scenario and Equation (6) is generated.
6. The final number of combined load-wind scenarios is obtained by multiplying the number of load scenarios by the number of wind scenarios. By multiplying the probability of the load scenario by the probability of wind speed scenario, the probability of the combined load-wind scenarios (π_s) can be calculated as follows [19]:

$$\pi_s = \pi_l \times \pi_w \qquad (7)$$

3. Problem Formulation

As mentioned earlier, a wide range of objective functions for the RPP in power systems can be represented. This matter enormously affects the control variables, state variables, and all constraints of the RPP problem. Thus, by a proper formulation, all objectives can be achieved, all constraints can be satisfied, and the feasibility of the problem can be ensured. Due to the fact that the probabilistic nature of the problem has a major impact on its formulation, it is very important to use the probabilistic variables accurately in the problem formulation.

3.1. Variables

The same as the other optimization problems in power systems, such as Optimal Power Flow (OPF), two types of variables, named control variables and state variables, are defined for the RPP.

Normally, for a typical RPP, control variables are defined as generators' voltage magnitudes, transformers tap settings, and output reactive power of VAR sources. Considering a scenario-based approach to model the uncertainty of the problem, the control variables set (U) for a probabilistic RPP are expressed as Equation (8) [14–17].

$$U = \begin{cases} V_{g_{i,s}}, & i \in \Omega_g, \, s \in \Omega_s \\ t_{k_{i,s}}, & i \in \Omega_{TapCh}, \, s \in \Omega_s \\ Q_{C_{i,s}}, & i \in \Omega_{Comp}, \, s \in \Omega_s \end{cases} \qquad (8)$$

where $V_{g_{i,s}}$ shows the voltage magnitude of the ith generator for the sth scenario, $t_{k_{i,s}}$ is used to assign the settings of the ith tap-changing transformer for the sth scenario, and $Q_{C_{i,s}}$ shows the output reactive power of the ith VAR compensator device for the sth scenario. Likewise, Ω_g, Ω_s, Ω_{TapCh}, and Ω_{Comp} symbolize the set of generators, set of scenarios, set of tap-changing transformers, and set of VAR compensator devices, respectively.

The state variables in a typical RPP consist of the generated active power by the slack bus, the generated reactive power by each of the existing generators, the voltage magnitude of the load buses, and the flow of the transmission lines.

Using a scenario-based approach to model the uncertainty of the problem, the state variables set (X) for a probabilistic RPP are expressed as Equation (9) [14–17].

$$
X = \begin{cases}
P_{G_{Slack,s}}, & s \in \Omega_s \\
Q_{G_{i,s}}, & i \in \Omega_g,\ s \in \Omega_s \\
V_{L_{i,s}}, & i \in \Omega_{PQ},\ s \in \Omega_s \\
S_{l,s}^{From}, & l \in \Omega_{Lines},\ s \in \Omega_s \\
S_{l,s}^{To}, & l \in \Omega_{Lines},\ s \in \Omega_s
\end{cases} \tag{9}
$$

where $P_{G_{Slack,s}}$ indicates the generated active power by the slack generator (bus) for the sth scenario, $Q_{G_{i,s}}$ is used to denote the generated reactive power by the ith generator for the sth scenario, $V_{L_{i,s}}$ shows the voltage magnitude of the ith load bus for the sth scenario, and $S_{l,s}^{From}$ and $S_{l,s}^{To}$ show the apparent power flow of the sending and receiving ends of the lth line for the sth scenario, respectively. Additionally, Ω_{PQ} and Ω_{Lines} specify the set of the load buses and the set of transmission lines, respectively.

3.2. Objective Functions

For probabilistic multi-objective RPP, the aim is to satisfy three main objectives. These objectives include the minimization of total VAR investment cost, minimization of voltage stability index (*L*-index), and maximization of loadability factor, which lead to a reduction in total active power losses and improvement of voltage stability.

3.2.1. Minimization of Total VAR Investment Cost

One of the important objectives in the RPP is the total cost of VAR planning. In spite of allocating the optimal location and capacity for VAR sources, optimal VAR planning can handle the RPP problem from economic aspects. For this reason, the first objective function is a cost-based objective function comprising two main parts, as follows:

(1). The first part evaluates the expected cost of energy loss (W_c) during the generated scenarios and is expressed as follows [16–26]:

$$
W_c = \pi_s \left(h \sum_{s \in \Omega_s} t_s P_{loss,s} \right) \tag{10}
$$

where $P_{loss,s}$ shows the active power losses during the sth scenario, t_s represents the duration of the sth scenario, h is a constant parameter that is related to the first part cost-based objective function and identifies the per-unit energy cost, and π_s denotes the probability of the sth scenario. To calculate the total active power losses, Equation (11) can be used as follows [27–30]:

$$
P_{loss,s} = \sum_{\substack{l \in \Omega_{Line} \\ l = (i,j)}} G_{(l,s)} \left(V_{i,s}^2 + V_{j,s}^2 - 2 V_{i,s} V_{j,s} \cos\left(\theta_{i,s} - \theta_{j,s}\right) \right) \tag{11}
$$

where $V_{i,s}$ and $V_{j,s}$ are the sending and receiving ends voltage magnitude of the lth transmission line for the sth scenario, respectively, $\theta_{i,s}$ and $\theta_{j,s}$ are the sending and receiving ends voltage angles of the lth transmission line for the sth scenario, respectively, and $G_{(l,s)}$ is used to designate the conductance of the lth transmission line for the sth scenario.

(2). The second part measures the expected cost of VAR investment (I_c) during the generated scenarios and is derived as follows [14–17]:

$$I_c = \sum_{s \in \Omega_s} \pi_s \Big(\sum_{i \in \Omega_{Comp}} (e_i + C_{Ci} Q_{Ci,s}) \Big) \tag{12}$$

where e_i and C_{Ci} are the fixed and variable installation costs of VAR sources, respectively.

Accordingly, the first objective function (f_1) can be derived as follows:

$$f_1 = F_c = W_c + I_c \tag{13}$$

where F_c shows the expected Total VAR Cost (TVC).

3.2.2. Minimization of Voltage Stability Index

In this paper, the L-index is proposed as the voltage stability index that is a well-known static voltage index [31]. In order to estimate the static voltage stability of the power system, the L-index should be calculated for all load buses (PQ buses). All the load buses that have higher values of the L-index than others are considered as the weak buses. Weak buses mostly suffer from a lack of reactive power and are prone to the voltage collapse. Equation (14) can be used to calculate the L-index (L_j) for the jth load bus, as follows [31]:

$$L_j = \left| 1 - \sum_{i=1}^{\Omega_g} F_{ji} \frac{\overline{V_i}}{\overline{V_j}} \right|, \qquad \forall j \in \Omega_{PQ} \tag{14}$$

where $\overline{V_i}$ shows the voltage of the ith generator, and $\overline{V_j}$ represents the voltage of the jth load bus. F_{ji} can be derived from Y_{bus} matrix of the system. Thus, by rearranging the current and voltage equations in power systems, as shown in Equation (15), the consecutive Y_{bus} matrix is achieved. Thereafter, using the arrays of the consecutive Y_{bus} matrix, the F_{ji} matrix can be calculated as Equation (16).

$$\begin{bmatrix} I_g \\ I_l \end{bmatrix} = \begin{bmatrix} Y_{gg} & Y_{gl} \\ Y_{lg} & Y_{ll} \end{bmatrix} \begin{bmatrix} V_g \\ V_l \end{bmatrix} \tag{15}$$

$$F_{ji} = -[Y_{ll}]^{-1} Y_{lg}, \qquad \forall j \in \Omega_{PQ}, \forall i \in \Omega_g \tag{16}$$

where I_g and I_l show the current of generators and loads, respectively, and V_g and V_l are the voltage of generators and loads, respectively. In addition, Y_{gg}, Y_{gl}, Y_{lg}, and Y_{ll} are the submatrices of the consecutive Y_{bus} matrix. It should be noted that only Y_{ll} arrays of the consecutive Y_{bus} matrix are related to the PQ nodes. Also, the consecutive Y_{bus} matrix is a symmetric matrix. Therefore, $Y_{lggl} = Y_{lg}$.

By minimizing the values of the L-index at the weak buses, there is a possibility to increase the level of static voltage stability in power systems. The voltage stability of power systems can be determined by the L-index when the maximum value of the L-index (L_{max}) is assigned to the static voltage stability level in power systems, as follows:

$$L_{max} = \max(L_j), \qquad \forall j \in \Omega_{PQ} \tag{17}$$

To improve the static voltage stability of power systems, it is necessary to minimize L_{max}. It should be noted that the equations proposed for the L-index are related to the deterministic problem. In the case of a probabilistic problem, considering all necessary modifications on the Y_{bus} matrix in each scenario, after re-formulating Equations (14)–(17), new equations can be rewritten as follows:

$$L_{j,s} = \left| 1 - \sum_{i=1}^{\Omega_g} F_{ji,s} \frac{\overline{V_{i,s}}}{\overline{V_{j,s}}} \right|, \qquad \forall j \in \Omega_{PQ}, \forall s \in \Omega_s \tag{18}$$

where $L_{j,s}$ indicates L-index value for the jth load bus and s^{th} scenario, $\overline{V_{i,s}}$ and $\overline{V_{j,s}}$ are the voltage of the ith generator and j^{th} load bus for the sth scenario, respectively. For each scenario, the $F_{ji,s}$ matrix can be calculated as Equation (20).

$$\begin{bmatrix} I_{g,s} \\ I_{l,s} \end{bmatrix} = \begin{bmatrix} Y_{gg,s} & Y_{gl,s} \\ Y_{lg,s} & Y_{ll,s} \end{bmatrix} \begin{bmatrix} V_{g,s} \\ V_{l,s} \end{bmatrix}, \qquad \forall s \in \Omega_s \tag{19}$$

$$F_{ji,s} = -\left[Y_{ll,s} \right]^{-1} Y_{lg,s}, \qquad \forall j \in \Omega_{PQ}, \forall i \in \Omega_g, \forall s \in \Omega_s \tag{20}$$

where $I_{g,s}$ and $I_{l,s}$ denote the current of generators and loads for the sth scenario, respectively, $V_{g,s}$ and $V_{l,s}$ are the voltage of generators and loads for the sth scenario, respectively. Also, $Y_{gg,s}$, $Y_{gl,s}$, $Y_{lg,s}$, and $Y_{ll,s}$ are the submatrices of the consecutive Y_{bus} matrix for the sth scenario.

According to the aforementioned descriptions, Equations (18)–(20) can be obtained for each scenario. The maximum value of the L-index for each scenario can be derived as follows:

$$L_{max,s} = \max\left(L_{j,s} \right), \qquad \forall j \in \Omega_{PQ}, \forall s \in \Omega_s \tag{21}$$

Consequently, the second objective function (f_2), which is the expected value of the static voltage stability index during the generated scenarios, can be derived as follows:

$$f_2 = \sum_{s \in \Omega_s} \pi_s L_{max,s} \tag{22}$$

3.2.3. Maximization of the Loadability Factor

The injected active power (P_i) and reactive power (Q_i) at the ith bus can be expressed in terms of the voltage (V), the elements of the Y_{bus} matrix of the system, and the loadability factor (Γ) as follows [32]:

$$P_i = P_{Gi} - (1 + \Gamma)P_{Di} - \mathrm{Re}\{V_i \sum_{j=1}^{N_B} \left(V_j Y_{i,j} \right)^* \} \tag{23}$$

$$Q_i = Q_{Gi} - (1 + \Gamma)Q_{Di} - \mathrm{Im}\{V_i \sum_{j=1}^{N_B} \left(V_j Y_{i,j} \right)^* \} \tag{24}$$

where P_{Gi} and Q_{Gi} are the active and reactive power generation at the ith bus, respectively, P_{Di} and Q_{Di} represent the base-case active and reactive power consumption at the ith bus, respectively, and N_B denotes the total number of buses.

Maximizing the loadability factor is defined as the third objective function, in this paper. However, considering the random nature of the problem, a probabilistic formulation is required. Therefore, by re-formulating Equations (23) and (24), a stochastic formula is derived to obtain the expected loadability factor, as shown in Equations (25) and (26).

$$P_{i,s} = P_{Gi,s} - (1 + \Gamma(s))P_{Di,s} - \mathrm{Re}\{V_{i,s} \sum_{j=1}^{N_B} \left(V_{j,s} Y_{i,j,s} \right)^* \}, \qquad \forall s \in \Omega_s \tag{25}$$

$$Q_{i,s} = Q_{Gi,s} - (1 + \Gamma(s))Q_{Di,s} - \text{Im}\{V_{i,s} \sum_{j=1}^{N_B} \left(V_{j,s} Y_{i,j,s}\right)^*\}, \qquad \forall s \in \Omega_s \tag{26}$$

where $P_{i,s}$ and $Q_{i,s}$ denote the injected active and reactive power at the ith bus for the sth scenario, respectively, $P_{Gi,s}$ and $Q_{Gi,s}$ represent the active and reactive power generation at the ith bus for the sth scenario, respectively, $P_{Di,s}$ and $Q_{Di,s}$ are the base-case active and reactive power consumption at the ith bus for the sth scenario, respectively, $\Gamma(s)$ denotes the loadability factor for the sth scenario; $V_{i,s}$ and $V_{j,s}$ indicate the voltage of the ith bus j^{th} bus for the sth scenario, respectively, and lastly, the elements of the Y_{bus} matrix for the sth scenario are shown by $Y_{i,j,s}$.

According to the above-mentioned descriptions, the third objective function (f_3), which is the expected value of the loadability factor, can be derived as follows:

$$f_3 = \sum_{s \in \Omega_s} \pi_s \Gamma(s) \tag{27}$$

Finally, the optimization criteria subjected to equality and inequality constraints are as follows:

$$\text{Optimization Criteria} = \begin{cases} \min(f_1) \\ \min(f_2) \\ \max(f_3) \end{cases} \tag{28}$$

3.3. Constraints

The role of constraints in creating a feasible space for the problem and satisfying optimality conditions to find optimal solutions is undeniable. For this reason, the correct expression of constraints is one of the major priorities in the problem formulation.

3.3.1. Equality Constraints

The power flow equations are taken as the equality constraints for the RPP. Using the output power of wind farms and also considering the probabilistic nature of the problem, Equations (25) and (26) can be rewritten as follows:

$$P_{i,s} = P_{Gi,s} + P_{Wi,s} - (1 + \Gamma(s))P_{Di,s} - \text{Re}\{V_{i,s} \sum_{j=1}^{N_B} \left(V_{j,s} Y_{i,j,s}\right)^*\}, \qquad \forall s \in \Omega_s \tag{29}$$

$$Q_{i,s} = Q_{Gi,s} + Q_{Wi,s} - (1 + \Gamma(s))Q_{Di,s} - \text{Im}\{V_{i,s} \sum_{j=1}^{N_B} \left(V_{j,s} Y_{i,j,s}\right)^*\}, \qquad \forall s \in \Omega_s \tag{30}$$

where $P_{Wi,s}$ and $Q_{Wi,s}$ show the output active and reactive power of the ith wind farm for the sth scenario, respectively. It should be noted that the output reactive power of the wind farms is neglected in this paper.

3.3.2. Inequality Constraints

To keep both the control and state variables within their specific limits, another set of constraints is added to the problem, named as inequality constraints. Those constraints involve Equations (31)–(38) [24,25].

● Limits on the Control Variables

The upper limit (V_{gi}^{max}) and lower limit (V_{gi}^{min}) of a generator voltage magnitude for the sth scenario can be applied, as follows:

$$V_{g_i}^{min} \leq V_{g_{i,s}} \leq V_{g_i}^{max}, \qquad \forall i \in \Omega_g, \forall s \in \Omega_s \tag{31}$$

For all tap-changing transformers in each scenario, the following constraint should be satisfied.

$$t_{k_i}^{min} \leq t_{k_{i,s}} \leq t_{k_i}^{max}, \qquad \forall i \in \Omega_{TapCh}, \forall s \in \Omega_s \tag{32}$$

where $t_{k_i}^{min}$ and $t_{k_i}^{max}$ show the minimum and maximum settings of the ith tap-changing transformer, respectively.

The output reactive power of the VAR sources in each scenario is as follows:

$$Q_{C_i}^{min} \leq Q_{C_{i,s}} \leq Q_{C_i}^{max}, \qquad \forall i \in \Omega_{Comp}, \forall s \in \Omega_s \tag{33}$$

where $Q_{C_i}^{min}$ and $Q_{C_i}^{max}$ show the minimum and maximum output reactive power of the ith VAR compensator device, respectively.

- Limits on the State Variables

In terms of the generation units, for each scenario, two important constraints should be satisfied; (1) the limitation on the generated active power of the slack bus and (2) the limitation on the generated reactive power of each generation unit. Those constraints are given as follows:

$$P_{G_{Slack}}^{min} \leq P_{G_{Slack,s}} \leq P_{G_{Slack}}^{max}, \qquad \forall s \in \Omega_s \tag{34}$$

$$Q_{G_i}^{min} \leq Q_{G_{i,s}} \leq Q_{G_i}^{max}, \qquad \forall i \in \Omega_g, \forall s \in \Omega_s \tag{35}$$

where $P_{G_{Slack}}^{min}$ and $P_{G_{Slack}}^{max}$ indicate the maximum and minimum generated active power of the slack bus for the sth scenario, respectively. In addition, $Q_{G_i}^{min}$ and $Q_{G_i}^{max}$ show the maximum and minimum generated reactive power of the ith generator for the sth scenario, respectively.

In order to prevent the voltage collapse or insulating problems, it is required to limit the voltage magnitude of loads for each scenario, as follows:

$$V_{L_i}^{min} \leq V_{L_{i,s}} \leq V_{L_i}^{max}, \qquad \forall i \in \Omega_{PQ}, \forall s \in \Omega_s \tag{36}$$

where $V_{L_i}^{min}$ and $V_{L_i}^{max}$ are considered as the lower and upper limits of the voltage magnitude at the ith load bus for the sth scenario, respectively.

To reduce the risk of overload in transmission lines, the apparent flow of the transmission lines should be lower than a specified value. Equations (37) and (38) enforce the apparent flow of transmission lines to be at the secure level, as follows:

$$S_{l,s}^{From} \leq S_l^{max}, \qquad \forall l \in \Omega_{Lines}, \forall s \in \Omega_s \tag{37}$$

$$S_{l,s}^{To} \leq S_l^{max}, \qquad \forall l \in \Omega_{Lines}, \forall s \in \Omega_s \tag{38}$$

where S_l^{max} indicates the maximum apparent flow of the lth transmission line.

3.4. Other Considerations in the Problem Formulation

There are other considerations in the problem formulation, which are listed as follows:

- The transformers tap settings and output reactive power of the VAR sources are treated as continuous variables. Therefore, the whole problem is stated as a probabilistic multi-objective nonlinear problem.

- Since the Y_{bus} matrix of power systems is dependent on the transformers tap settings and due to the fact that the transformers tap settings are defined as scenario-dependent variables, the Y_{bus} matrix should be calculated for each scenario separately.

- The L-index value varies between 0 and 1 for power systems. It should be noted that except for the defined boundaries, the L-index value should be obtained without any further restriction during the optimization procedure.

4. Optimization Method

Once an optimization problem is formulated carefully, it is required to solve the problem via an optimization method. To choose an optimization method to solve a problem, it is necessary to consider the number of optimization variables, the type of variables, the number of objective functions, the number of constraints, and the convexity or non-convexity of the problem and the other characteristics [27–29]. In this regard, the optimization methods can be classified into three major groups; (1) exact methods based on mathematical calculations, (2) heuristic methods, and (3) combination of the exact methods and heuristic methods.

4.1. Multi-Objective Optimization Using ε-Constraint Method

According to [33,34], the ε-constraint method is considered as one of the classic methods for multi-objective optimization. This method is in line with the exact methods. In addition to its efficiency and simplicity, this method is applicable to both convex and non-convex problems. The main idea of the ε-constraint method is to reformulate the multi-objective problem as a single-objective problem. Then, by iteratively solving the single-objective problem, a Pareto Front is obtained. In the following, the details of the ε-constraint method are explained [34].

Considering a multi-objective problem ($\Psi(X)$), as shown in Equation (39), subjected to different constraints that should be optimized, the following steps should be taken.

$$\Psi(X) = (f_1(X), f_2(X), \dots f_i(X)), \qquad i = 1, 2, \dots, n \tag{39}$$

where $f_i(X)$ denotes the ith objective function and n shows the maximum number of existing objective functions.

1. Each objective function ($f_i(X)$) is optimized with the existing constraints separately and the results are saved in a table, called the payoff table.

2. According to the priority of the objective functions, one objective function is selected as the main objective function. Then, the rest of the objective functions are treated as new constraints and added to the main constraints. It should be noted that except for the main objective function, if the goal is to minimize and maximize all the objective functions, then, $f_i(X) \le e_i$ and $f_i(X) \ge e_i$, respectively. Also, e_i is a variable parameter.

3. In order to assign values to e_i, the maximum (f_i^{max}) and minimum (f_i^{min}) values of each objective function should be considered, as shown in Equation (40). It should be noted that those values can be obtained from the payoff table.

$$f_i^{min} \le f_i(X) \le f_i^{max} \tag{40}$$

4. To generate different values for e_{i,n_i}, Equations (41) and (42) are used to minimize and maximize the objective function, respectively. By dividing the domain of the ith objective function into q_i equal parts using Equations (41) and (42), q_i different values are obtained for e_{i,n_i}. It should be noted that n_i denotes the number of available generated values for e_{i,n_i}.

$$e_{i,n_i} = f_i^{max} - (\frac{f_i^{max} - f_i^{min}}{q_i})n_i, \qquad n_i = 0, 1, \dots, q_i \tag{41}$$

$$e_{i,n_i} = f_i^{min} - \left(\frac{f_i^{max} - f_i^{min}}{q_i}\right)n_i, \qquad n_i = 0, 1, \ldots, q_i \tag{42}$$

5. By using the obtained values from Step 4, it can be derived that $f_i(X) \leq e_{i,n_i}$ or $f_i(X) \geq e_{i,n_i}$. For different values of e_i, a set of solutions is obtained, which forms the Pareto front of the problem.

According to the above-mentioned descriptions, to solve the probabilistic multi-objective RPP, the following equation is formed.

$$\min f_1 \tag{43}$$

subjected to

$$\begin{cases} f_2 \leq e_{2,n_2} \\ f_3 \leq e_{3,n_3} \\ \text{Equations (29)-(38)} \end{cases} \tag{44}$$

4.2. Fuzzy Decision Maker (FDM)

As already mentioned, after solving a multi-objective optimization problem, a set of optimal solutions is obtained, called the Pareto Front. While only one solution from the Pareto Front can be chosen as the final optimal solution to the problem, which is known as the Best Compromise Solution (BCS). One way to choose the BCS is to use the Fuzzy Decision Maker (FDM). Having used a fuzzy membership function, each of the optimal solutions is mapped between 0 and 1. For the kth objective function, F_k, the linear fuzzy membership is defined as Equation (45) [24], and it is supposed that all the objective functions are minimized.

$$\hat{F}_k = \begin{cases} 1, & F_k \leq F_k^{min} \\ \frac{F_k^{max} - F_k}{F_k^{min} - F_k^{max}}, & F_k^{min} \leq F_k \leq F_k^{max} \\ 0, & F_k \geq F_k^{max} \end{cases} \tag{45}$$

where $\hat{F}_k$ represents the kth normalized objective function. In addition, F_k^{min} and F_k^{max} are used to express the minimum and maximum values of the kth objective function, respectively.

After obtaining the fuzzy values of each objective function using Equation (45), there are several ways to find the BCS. In this paper, to obtain the BCS, the min-max method, which is introduced in [35], is used.

$$BCS = \max\left(\min\left(\hat{F}_1, \hat{F}_2, \ldots, \hat{F}_k\right)\right) \tag{46}$$

5. Simulation Results and Discussions

To evaluate the performance of the proposed ε-constraint method in the presence of various objectives, containing expected total VAR cost (f_1), expected active power losses, expected voltage stability index (f_2), and expected loadability factor (f_3), two deterministic and three probabilistic cases are studied, as follows:

A. Deterministic multi-objective RPP without considering the loadability factor (assessing the proficiency of ε-constraint method)
B. Deterministic multi-objective RPP considering the loadability factor
C. Probabilistic multi-objective RPP considering the load demand uncertainty
D. Probabilistic multi-objective RPP considering the wind power generation uncertainty
E. Probabilistic multi-objective RPP considering load demand and wind power generation uncertainties at the same time

All the cases are implemented in GAMS environment Ver. 25.1.2 [36–39], and are solved using the CONOPT 3 Solver [40], in an ASUS laptop, with 8 GB of RAM and 2.4 GHz. The descriptions of the case study are presented in the next subsection.

5.1. Case Study Descriptions and Simulation Results

The test system is the IEEE 30-bus test system, which has 6 generation units, 4 transformers, and 41 branches. The initial settings of the generators' voltage magnitude and transformers tap settings are obtained from [30]. Figure 1 shows the single line diagram of the IEEE-30-bus test system. Also, both the output active and reactive power of generators are set according to [41]. The loads' data and line data are available in [42]. It is assumed that there is not any VAR source in the case study.

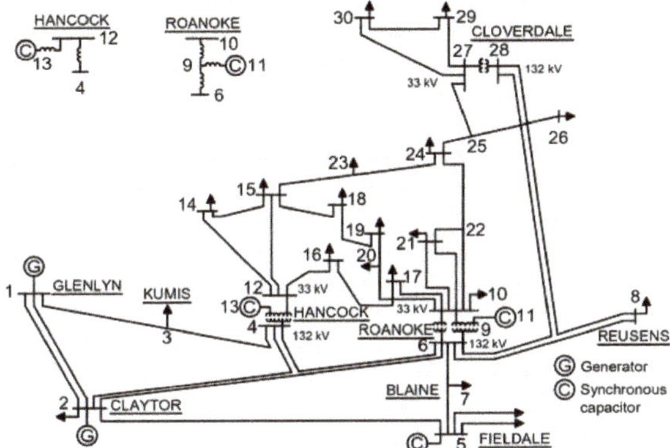

Figure 1. Single line diagram of the IEEE-30 bus test system [43].

To allocate appropriately the VAR compensation devices, firstly, the *L*-index should be determined for the load buses. Then, the load buses with high values of *L*-index are taken into account as the candidate buses for the VAR compensation devices installation. It is observed that after implementing the proposed methodology, the load at bus 24, bus 25, bus 26, bus 29, and bus 30 obtain the higher values of *L*-index than the other load buses. As a result, the VAR compensator buses are found. After the allocation of VAR compensation devices, it is supposed that the capacities of the VAR compensators can be set to zero. The initial conditions of the system considering the full-load and 1-year planning are stated in Table 2.

Table 2. Control variables and objectives under initial conditions.

Generator Voltage Magnitude	
V_{g1} (p.u.)	1.050
V_{g2} (p.u.)	1.044
V_{g5} (p.u.)	1.023
V_{g8} (p.u.)	1.025
V_{g11} (p.u.)	1.050
V_{g13} (p.u.)	1.050
Transformer Tap Settings	
t_{6-9} (p.u.)	0.950
t_{6-10} (p.u.)	1.100
t_{4-12} (p.u.)	1.025
t_{28-27} (p.u.)	1.050
VAR Compensator	
Q_{c24} (MVAR)	0.000
Q_{c25} (MVAR)	0.000
Q_{c26} (MVAR)	0.000
Q_{c29} (MVAR)	0.000
Q_{c30} (MVAR)	0.000
Objective	
P_{loss} (MW)	5.4970
f_1 ($)	2.8892×10^6
f_2	0.1635

The control variables limits are expressed in Table 3. The per-unit energy cost is equal to 0.06 (\$/h) [3], the fixed installation cost (e_i) for all VAR sources is 1000 (\$), and the variable installation cost (C_{Ci}) for all VAR sources is 3.0 (\$/kVAR) [44].

Table 3. Control variables limits.

Control Variable	Value
$V_{g_i}^{min}$ (p.u.)	0.900
$V_{g_i}^{max}$ (p.u.)	1.100
$t_{k_i}^{min}$ (p.u.)	0.900
$t_{k_i}^{max}$ (p.u.)	1.100
$Q_{C_i}^{min}$ (MVAR)	0.000
$Q_{C_i}^{max}$ (MVAR)	35.00

The voltage magnitude at the load buses, which are considered as the state variables, must be limited between 0.95 (p.u.) and 1.05 (p.u.). To show the effectiveness of the proposed method, two deterministic and three probabilistic cases are considered in the following subsections.

5.1.1. Case A: Deterministic Multi-Objective RPP without Considering the Loadability Factor

In order to validate the efficiency of the proposed method for multi-objective RPP, the ε-constraint method is applied to a deterministic multi-objective RPP problem. The obtained results are also compared with the approach presented in [3]. The deterministic multi-objective RPP aims to minimize the total VAR cost and voltage stability index. Thus, it is expected to achieve a reduction in active power losses and an improvement in the voltage stability index. Table 4 shows the obtained results from deterministic multi-objective RPP without considering the loadability factor in IEEE 30-bus test system with the initial settings. The duration of the load for deterministic multi-objective RPP is assumed to be 8760 h for full-load condition and without changes in the load level. According to Table 4, 15 Pareto optimal solutions are obtained by the ε-constraint method. After that, the min-max approach chooses the fifth solution (highlighted row) as the BCS. The active power losses for the BCS are 4.9813 MW.

Table 4. Obtained Pareto optimal solutions for the deterministic multi-objective RPP without considering the loadability factor.

x	f_1 (\$)	f_2	$\hat{F}_1$	$\hat{F}_2$	$\min(\hat{F}_1,\hat{F}_2)$
1	3.0334×10^6	0.1241	0.0000	1.0000	0.0000
2	2.9064×10^6	0.1241	0.3033	0.9286	0.3033
3	2.8182×10^6	0.1246	0.5139	0.8571	0.5139
4	2.7532×10^6	0.1249	0.6692	0.7857	0.6692
5	2.7084×10^6	0.1252	0.7762	0.7143	0.7143
6	2.6802×10^6	0.1255	0.8437	0.6429	0.6429
7	2.6611×10^6	0.1257	0.8892	0.5714	0.5714
8	2.6496×10^6	0.1260	0.9166	0.5000	0.5000
9	2.6415×10^6	0.1263	0.9360	0.4286	0.4286
10	2.6350×10^6	0.1266	0.9516	0.3571	0.3571
11	2.6294×10^6	0.1269	0.9649	0.2857	0.2857
12	2.6247×10^6	0.1271	0.9762	0.2143	0.2143
13	2.6207×10^6	0.1274	0.9858	0.1429	0.1429
14	2.6174×10^6	0.1277	0.9936	0.0714	0.0714
15	2.6147×10^6	0.1280	1.0000	0.0000	0.0000

Considering the same operating condition, the ε-constraint method is compared with the Multi-Objective Differential Evolution (MODE) algorithm, which is recommended to solve the deterministic multi-objective RPP [3]. For the BCS, the results of the comparison are presented in Table 5. As it can be observed from Table 5, the ε-constraint method shows better performance compared with the MODE algorithm in minimizing total VAR cost and active power losses. The superiority of the ε-constraint method is confirmed by a 6.2578 % reduction in total VAR cost and a 9.3815 % decrease in the active power losses over the Base Case. However, as it can be observed from Table 5, the voltage stability index of the conventional method is better than the proposed approach.

Table 5. Comparison of the obtained results between the ε-constraint method and MODE algorithm for the BCS under the same operating conditions.

Method		P_{loss} (MW)	f_1($)	f_2
ε-constraint Method	Base Case	5.4970	2.8892×10^6	0.16350
	BCS	4.9813	2.7084×10^6	0.12520
	Reduction (%)	9.3815	6.2578	23.4251
MODE Algorithm	Base Case	4.9630	2.6085×10^6	0.19780
	BCS	4.8300	2.5387×10^6	0.12040
	Reduction (%)	2.6798	2.6759	39.1304

The optimal values of the control variables for Case A are represented in Table 6. As it can be observed, only one VAR source has a value of zero. Note that the fixed installation VAR cost is also considered for all VAR sources with the value of zero during the planning studies. It is apparent from Case A that the ε-constraint is an effective method to generate Pareto optimal solutions for the multi-objective RPP.

Table 6. The optimal values of the control variables for the deterministic multi-objective RPP without considering the loadability factor.

Control Variable	Optimal Value
V_{g1} (p.u.)	1.06940
V_{g2} (p.u.)	1.06150
V_{g5} (p.u.)	1.04110
V_{g8} (p.u.)	1.04260
V_{g11} (p.u.)	1.10000
V_{g13} (p.u.)	1.05550
t_{6-9} (p.u.)	1.03640
t_{6-10} (p.u.)	0.92960
t_{4-12} (p.u.)	0.97700
t_{28-27} (p.u.)	0.99910
Q_{c24} (MVAR)	20.9529
Q_{c25} (MVAR)	1.84190
Q_{c26} (MVAR)	2.34290
Q_{c29} (MVAR)	3.28580
Q_{c30} (MVAR)	0.00000

5.1.2. Case B: Deterministic Multi-Objective RPP Considering the Loadability Factor

In this part, the ε-constraint method is applied to the multi-objective RPP problem in a complex form. However, the uncertainties of the load demand and wind power generation are not considered in this case. In comparison with Case A, another objective, which is called the loadability factor, is added to the problem. Therefore, the main objectives in this part include minimizing the total VAR cost, reducing the active power losses, improving the voltage stability index, and maximizing the loadability factor. In order to solve a deterministic multi-objective RPP considering the loadability factor, it is assumed that the system is under full-load condition. The duration of the load is assumed to be 8760 h.

Table 7 provides the simulation results of the deterministic multi-objective RPP problem considering the loadability factor. In addition, this table illustrates that among the 15 generated Pareto optimal solutions, the eighth solution (highlighted row) is the BCS through the min-max approach. The active power losses for the BCS is 9.3494 MW. Also, it can be observed that the total VAR cost and active power losses are dramatically increased for BCS in comparison with Case A. Moreover, the voltage stability index is not improved compared with Case A. Nevertheless, the loadability factor is improved in Case B. The main reason behind the deterioration of active power losses and voltage stability index is due to the enhancement of the loadability factor. It should be noted that the loadability factor can hugely affect the active power losses and voltage stability of power systems.

Table 7. Obtained Pareto optimal solutions for the deterministic multi-objective RPP considering the loadability factor.

x	f_1 (\$)	f_2	f_3	$\hat{F}_1$	$\hat{F}_2$	$\hat{F}_3$	$\min(\hat{F}_1,\hat{F}_2,\hat{F}_3)$
1	2.9015×10^6	0.1272	0.0000	1.0000	1.0000	0.0000	0.0000
2	3.1145×10^6	0.1309	0.0236	0.9511	0.9286	0.0724	0.0724
3	3.3885×10^6	0.1347	0.0472	0.8881	0.8571	0.1448	0.1448
4	3.7008×10^6	0.1384	0.0708	0.8164	0.7857	0.2172	0.2172
5	4.0028×10^6	0.1421	0.0944	0.7470	0.7143	0.2896	0.2896
6	4.3051×10^6	0.1459	0.1181	0.6776	0.6429	0.3620	0.3620
7	4.6107×10^6	0.1496	0.1417	0.6074	0.5714	0.4344	0.4344
8	4.9361×10^6	0.1533	0.1653	0.5326	0.5000	0.5068	0.5000
9	5.2798×10^6	0.1571	0.1889	0.4537	0.4286	0.5792	0.4286
10	5.6345×10^6	0.1608	0.2125	0.3722	0.3571	0.6515	0.3571
11	5.9691×10^6	0.1645	0.2361	0.2953	0.2857	0.7239	0.2857
12	6.2864×10^6	0.1683	0.2597	0.2224	0.2143	0.7963	0.2143
13	6.6184×10^6	0.1720	0.2833	0.1462	0.1429	0.8687	0.1429
14	6.9648×10^6	0.1757	0.3070	0.0666	0.0714	0.9411	0.0666
15	7.2547×10^6	0.1795	0.3262	0.0000	0.0000	1.0000	0.0000

Table 8 depicts the optimal values of the control variables for Case B. As it can be observed, three VAR sources have a value of zero. It should be noted that the fixed installation VAR cost is also considered for all VAR sources with the value of zero during the planning studies.

Table 8. The optimal values of the control variables for the deterministic multi-objective RPP considering the loadability factor.

Control Variable	Optimal Value
V_{g1} (p.u.)	1.06300
V_{g2} (p.u.)	1.05310
V_{g5} (p.u.)	1.07610
V_{g8} (p.u.)	1.05420
V_{g11} (p.u.)	1.10000
V_{g13} (p.u.)	1.08520
t_{6-9} (p.u.)	1.02690
t_{6-10} (p.u.)	0.90000
t_{4-12} (p.u.)	1.00600
t_{28-27} (p.u.)	0.97310
Q_{c24} (MVAR)	0.00000
Q_{c25} (MVAR)	0.00000
Q_{c26} (MVAR)	0.00000
Q_{c29} (MVAR)	3.03920
Q_{c30} (MVAR)	2.66360

5.1.3. Case C: Probabilistic Multi-Objective RPP Considering the Load Demand Uncertainty

In Cases A and B, the multi-objective RPP in power systems is solved using the deterministic approach. However, with the increasing level of uncertainty, probabilistic multi-objective is required for the RPP problem. In Case C, the probabilistic multi-objective RPP considering three different scenarios for the load level is performed. Each scenario for the load level consists of two main parts: (1) probability of the load level and (2) duration of the load. The overall duration of the load is assumed to be 8760 h, which is the expected time horizon for the RPP. The specifications of the system loading are described in Table 9.

Table 9. The specifications of the system loading for probabilistic multi-objective RPP considering the load demand uncertainty.

Scenario	Level of the Load	Probability	Duration of the Load (h)
S_1	0.95	0.1	2920
S_2	1.00	0.8	4380
S_3	1.05	0.1	1460

The simulation results obtained from the probabilistic multi-objective RPP considering the load demand uncertainty are given in Table 10. As it can be observed from this table, 15 Pareto optimal solutions are generated using the ε-constraint method. Using the min-max approach, the eighth solution (highlighted row) is chosen as the BSC. It is worth mentioning that the expected active power losses are 9.5049 MW for the BCS. From Table 10, it is clear that the expected total VAR cost is reduced compared with the Base Case. The expected voltage stability index and the expected loadability factor also show improvement towards the initial conditions. However, with more considerations, it is revealed that the expected active power losses are increased. This fact stems from the evident increase in the loadability factor. As a common incidence in power systems, following the escalation of the loadability factor, the active power losses increase and the system becomes voltage unstable. Generally, from the power systems operators' perspective, monitoring of the voltage magnitude at the load buses as a way of preventing voltage collapse is in high priority. Therefore, the voltage profile of the load buses for each loading scenario is plotted for the BCS, as shown in Figure 2.

Table 10. Obtained Pareto optimal solutions for the probabilistic multi-objective RPP considering the load demand uncertainty.

x	f_1 ($\$$)	f_2	f_3	$\hat{F}_1$	$\hat{F}_2$	$\hat{F}_3$	$\min(\hat{F}_1,\hat{F}_2,\hat{F}_3)$
1	1.3129×10^6	0.1294	0.0000	1.0000	1.0000	0.0000	0.0000
2	1.4223×10^6	0.1339	0.0237	0.9443	0.9197	0.0722	0.0722
3	1.5418×10^6	0.1383	0.0473	0.8834	0.8395	0.1443	0.1443
4	1.6647×10^6	0.1428	0.0710	0.8207	0.7592	0.2165	0.2165
5	1.7954×10^6	0.1472	0.0946	0.7542	0.6790	0.2887	0.2887
6	1.9328×10^6	0.1517	0.1183	0.6841	0.5987	0.3609	0.3609
7	2.0774×10^6	0.1561	0.1420	0.6105	0.5185	0.4330	0.4330
8	2.2510×10^6	0.1606	0.1688	0.5220	0.4382	0.5148	0.4382
9	2.3848×10^6	0.1637	0.1893	0.4538	0.3818	0.5774	0.3818
10	2.5365×10^6	0.1673	0.2131	0.3765	0.3172	0.6501	0.3172
11	2.7000×10^6	0.1713	0.2379	0.2932	0.2447	0.7256	0.2447
12	2.9942×10^6	0.1783	0.2852	0.1433	0.1187	0.8700	0.1187
13	3.1909×10^6	0.1828	0.3143	0.0431	0.0376	0.9586	0.0376
14	3.2627×10^6	0.1844	0.3252	0.0065	0.0072	0.9919	0.0065
15	3.2754×10^6	0.1848	0.3278	0. 0000	0. 0000	1. 000	0.0000

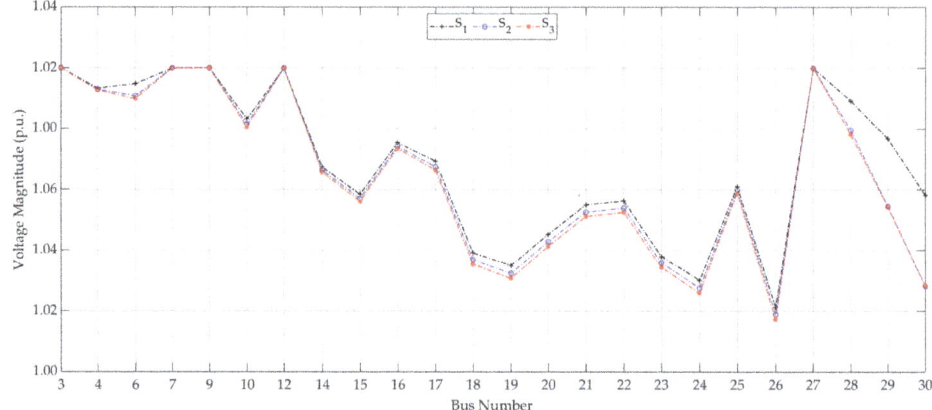

Figure 2. Voltage profile of the load buses for probabilistic multi-objective RPP considering the load demand uncertainty.

The optimal values of the control variables for the BSC among the different load scenarios are given in Table 11. Having checked this table closely, it is noticed that some VAR sources have a value of zero. Consequently, the variable cost of those VAR sources is equal to zero. However, those VAR sources are allocated, and their fixed VAR installation costs are considered for the planning studies in this paper.

Table 11. The optimal values of the control variables among the load scenarios for probabilistic multi-objective RPP considering the load demand uncertainty.

Control Variable	S_1	S_2	S_3	Expected Value
V_{g1} (p.u.)	1.0621	1.0632	1.0640	1.0632
V_{g2} (p.u.)	1.0530	1.0532	1.0533	1.0532
V_{g5} (p.u.)	1.0749	1.0783	1.0794	1.0780
V_{g8} (p.u.)	1.0550	1.0522	1.0514	1.0524
V_{g11} (p.u.)	1.1000	1.1000	1.1000	1.1000
V_{g13} (p.u.)	1.0828	1.0903	1.0927	1.0898
t_{6-9} (p.u.)	1.0285	1.0249	1.0234	1.0251
t_{6-10} (p.u.)	0.9000	0.9000	0.9000	0.9000
t_{4-12} (p.u.)	1.0036	1.0143	1.0171	1.0135
t_{28-27} (p.u.)	0.9721	0.9500	0.9492	0.9521
Q_{c24} (MVAR)	0.0000	0.0000	0.0000	0.0000
Q_{c25} (MVAR)	0.0000	0.0000	0.0000	0.0000
Q_{c26} (MVAR)	0.0000	0.0000	0.0000	0.0000
Q_{c29} (MVAR)	1.8307	0.0000	0.0000	0.1831
Q_{c30} (MVAR)	3.1448	0.0000	0.1993	0.3344

It is clear from Figure 2 that the voltage magnitude of the load buses remains in the range of 0.95 p.u. and 1.05 p.u. for all three load scenarios. Although the loadability factor is improved, the voltage stability index of the system is ensured from the voltage magnitude point of view. Therefore, by making an allowance for the load demand uncertainty, the obtained total VAR cost seems to be more realistic. In the same way, the voltage stability index and the loadability factor are more reliable due to including more scenarios for the planning horizon.

5.1.4. Case D: Probabilistic Multi-Objective RPP Considering the Wind Power Generation Uncertainty

In Case D, it is assumed that a wind farm is located at a PQ node. After generating the wind speed scenarios using the Weibull distribution and power curve, a probabilistic multi-objective RPP is performed. In this case, the IEEE 30-bus test system is modified based on [45]. Hence, a wind farm with a rated power of 40 MW is added to bus 22. The wind farm data is derived from [45] and is presented in Appendix A. To evaluate the impact of the wind farm, six scenarios for the output power of the wind farm are generated. The duration of the load is assumed to be 1460 h and without changes in the load level. The generated wind scenarios and their details are given in Table 12.

Table 12. Generated wind scenarios for the probabilistic multi-objective RPP considering the wind power generation uncertainty.

Scenario	Wind Power Generation (MW)	Probability	Level of the Load	Duration of the Load (h)
S_1	0.00000	0.0861	1	1460
S_2	5.27050	0.1212	1	1460
S_3	15.0917	0.1492	1	1460
S_4	24.9726	0.1546	1	1460
S_5	34.8784	0.1413	1	1460
S_6	40.0000	0.3476	1	1460

The obtained results from the probabilistic multi-objective RPP using the generated wind scenarios are represented in Table 13. As it is observed from this table, among the 15 generated Pareto optimal solutions using the ε-constraint technique, the eighth solution (highlighted row) is selected as the BSC after applying the min-max approach. The corresponding value of expected active power losses is 8.5777 MW. Compared with the Base Case and Case A, it is clear that the expected total VAR cost has had a remarkable reduction for the best compromise solution. In addition, the enhancement of expected voltage stability index and expected loadability factor is undeniable towards the Base Case and Case A. In addition, the expected active power losses are elevated in contrast with the Base Case. However, the expected active power losses show a reduction of roughly 1 MW, when it is compared with Case A. The main reason for this reduction is the existence of the wind farm in the case study.

Table 13. Obtained Pareto optimal solutions for the probabilistic multi-objective RPP considering the wind generation uncertainty.

x	$f_1(\$)$	f_2	f_3	$\hat{F}_1$	$\hat{F}_2$	$\hat{F}_3$	$\min(\hat{F}_1,\hat{F}_2,\hat{F}_3)$
1	3.7924×10^5	0.1222	0.0000	1.0000	1.0000	0.0000	0.0000
2	4.1835×10^5	0.1282	0.0300	0.9538	0.9096	0.0724	0.0724
3	4.6485×10^5	0.1335	0.0600	0.8988	0.8293	0.1448	0.1448
4	5.1663×10^5	0.1386	0.0900	0.8376	0.7523	0.2172	0.2172
5	5.7321×10^5	0.1442	0.1200	0.7707	0.6683	0.2896	0.2896
6	6.3071×10^5	0.1480	0.1500	0.7028	0.6093	0.3620	0.3620
7	6.9212×10^5	0.1525	0.1800	0.6302	0.5418	0.4344	0.4344
8	7.5741×10^5	0.1570	0.2100	0.5530	0.4740	0.5068	0.4740
9	8.2287×10^5	0.1615	0.2400	0.4757	0.4058	0.5792	0.4058
10	8.8964×10^5	0.1661	0.2700	0.3968	0.3355	0.6515	0.3355
11	9.4995×10^5	0.1706	0.3000	0.3255	0.2680	0.7239	0.2680
12	1.0169×10^6	0.1752	0.3300	0.2463	0.1984	0.7963	0.1984
13	1.0876×10^6	0.1798	0.3600	0.1628	0.1285	0.8687	0.1285
14	1.1621×10^6	0.1845	0.3900	0.0747	0.0583	0.9411	0.0583
15	1.2253×10^6	0.1883	0.4144	0.0000	0.0000	1.0000	0.0000

The optimal values of the control variables for the BCS over the wind scenarios are represented in Table 14. Taking a look at Table 14, it is shown that the VAR sources gain the value of zero in almost all scenarios. Therefore, the variable VAR investment cost for those VAR sources equals to zero. However, the VAR sources are allocated and their fixed VAR investment costs are taken into account during the planning horizon.

Table 14. The optimal values of the control variables among the wind scenarios for probabilistic multi-objective RPP considering the wind power generation uncertainty.

Control Variable	S_1	S_2	S_3	S_4	S_5	S_6	Expected Value
V_{g1} (p.u.)	1.0630	1.0627	1.0619	1.0611	1.0604	1.0599	1.0611
V_{g2} (p.u.)	1.0531	1.0530	1.0527	1.0523	1.0520	1.0518	1.0523
V_{g5} (p.u.)	1.0767	1.0778	1.0773	1.0769	1.0766	1.0765	1.0769
V_{g8} (p.u.)	1.0536	1.0525	1.0528	1.0531	1.0533	1.0534	1.0531
V_{g11} (p.u.)	1.1000	1.1000	1.1000	1.1000	1.1000	1.1000	1.1000
V_{g13} (p.u.)	1.0867	1.0893	1.0887	1.0883	1.0882	1.0883	1.0883
t_{6-9} (p.u.)	1.0264	1.0266	1.0289	1.0311	1.0330	1.0338	1.0310
t_{6-10} (p.u.)	0.9000	0.9000	0.9000	0.9000	0.9000	0.9000	0.9000
t_{4-12} (p.u.)	1.0084	1.0121	1.0094	1.0071	1.0052	1.0043	1.0070
t_{28-27} (p.u.)	0.9657	0.9499	0.9490	0.9480	0.9468	0.9462	0.9491
Q_{c24} (MVAR)	0.0000	0.0000	0.0000	0.0000	0.0000	0.0000	0.0000
Q_{c25} (MVAR)	0.0000	0.0000	0.0000	0.0000	0.0000	0.0000	0.0000
Q_{c26} (MVAR)	0.0000	0.0000	0.0000	0.0000	0.0000	0.0000	0.0000
Q_{c29} (MVAR)	0.0000	0.0000	0.0000	0.0000	0.0000	0.0000	0.0000
Q_{c30} (MVAR)	3.8586	0.0000	0.0000	0.0000	0.0000	0.0000	0.3321

The voltage profile of the load buses for the BCS is plotted in Figure 3. As it can be observed, the voltage magnitude of the load buses is kept in the interval of 0.95 p.u. and 1.05 p.u. during all wind scenarios. Hence, it can be concluded that based on a proper RPP and having adequate reactive power reserve, the voltage magnitude of the load buses are restricted with specific limits.

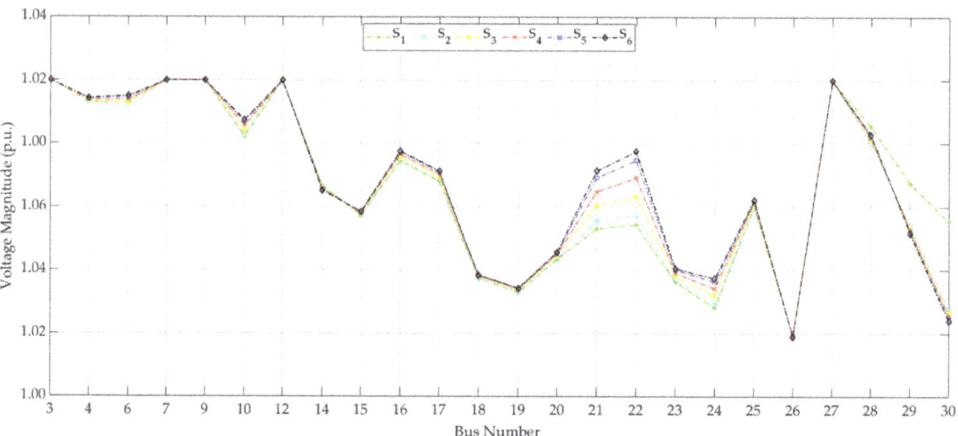

Figure 3. Voltage profile of the load buses for probabilistic multi-objective RPP considering the wind power generation uncertainty.

5.1.5. Case E: Probabilistic Multi-Objective RPP Considering Load Demand and Wind Power Generation Uncertainties

To have a more comprehensive overlook of the probabilistic RPP, in Case E, it is preferred to perform RPP in the presence of two stochastic input variables, including the load demand and wind power generation. Hence, considering the level of the load and wind power generation as the stochastic input variables, combined load-wind scenarios are generated via the proposed technique. After determining the load-wind scenarios, a probabilistic multi-objective RPP is performed to evaluate the existing objectives, while the number of random input variables increases. Taking the IEEE 30 bus-test system with initial settings as the benchmark, 18 combined load-wind scenarios are generated. In general, three load scenarios and six wind scenarios are used to generate 18 combined load-wind scenarios. The descriptions of generated load-wind scenarios are given in Table 15.

Table 15. Generated load-wind scenarios for probabilistic multi-objective RPP considering load demand and wind power generation uncertainties.

Scenario	Wind Power Generation (MW)	Level of the Load	Duration of the Load (h)	Probability
S_1	0.00000	0.95	400	0.0086
S_2	5.27050	0.95	400	0.0121
S_3	15.0917	0.95	400	0.0149
S_4	24.9726	0.95	400	0.0155
S_5	34.8784	0.95	400	0.0141
S_6	40.0000	0.95	400	0.0348
S_7	0.00000	1.00	730	0.0689
S_8	5.27050	1.00	730	0.0970
S_9	15.0917	1.00	730	0.1194
S_{10}	24.9726	1.00	730	0.1237
S_{11}	34.8784	1.00	730	0.1130
S_{12}	40.0000	1.00	730	0.2781
S_{13}	0.00000	1.05	330	0.0086
S_{14}	5.27050	1.05	330	0.0121
S_{15}	15.0917	1.05	330	0.0149
S_{16}	24.9726	1.05	330	0.0155
S_{17}	34.8784	1.05	330	0.0141
S_{18}	40.0000	1.05	330	0.0348

Table 16 shows the obtained results for the probabilistic multi-objective RPP using the generated load-wind scenarios. As seen from Table 16, by applying the min-max method among the 15 generated Pareto optimal solutions using the ε-constraint approach, the eighth solution (highlighted row) is selected as the BCS. The associated value to expected active power losses is 8.5575 MW. It can be observed that the expected Total VAR cost is considerably reduced, while the expected voltage stability index and the expected loadability factor are not significantly improved towards Case B. In contrast with Case A and the Base Case, both the expected voltage stability index and the expected loadability factor are improved compared with case B. Considering the expected active power losses, no substantial decrease is observed in Case C when it is compared with Case B. Compared with Cases A and C, a reduction of about 1 MW in expected active power losses can be estimated. Due to enhancing the expected loadability factor, the expected active power losses escalate relative to the Base Case.

Table 16. Obtained Pareto optimal solutions for the probabilistic multi-objective RPP considering load demand and wind power generation uncertainties.

χ	$f_1(\$)$	f_2	f_3	$\hat{F}_1$	$\hat{F}_2$	$\hat{F}_3$	$\min(\hat{F}_1, \hat{F}_2, \hat{F}_3)$
1	2.0134×10^5	0.1172	0.0011	1.0000	1.0000	0.0000	0.0000
2	2.1190×10^5	0.1224	0.0307	0.9718	0.9277	0.0717	0.0717
3	2.2563×10^5	0.1277	0.0607	0.9352	0.8555	0.1441	0.1441
4	2.4456×10^5	0.1330	0.0907	0.8847	0.7832	0.2165	0.2165
5	2.6724×10^5	0.1382	0.1207	0.8242	0.7110	0.2889	0.2889
6	2.9247×10^5	0.1435	0.1507	0.7569	0.6387	0.3614	0.3614
7	3.1866×10^5	0.1487	0.1807	0.6871	0.5667	0.4338	0.4338
8	3.4683×10^5	0.1539	0.2107	0.6119	0.4946	0.5062	0.4946
9	3.7578×10^5	0.1592	0.2407	0.5347	0.4224	0.5786	0.4224
10	4.0375×10^5	0.1644	0.2707	0.4601	0.3503	0.6511	0.3503
11	4.3143×10^5	0.1697	0.3007	0.3863	0.2782	0.7235	0.2782
12	4.6080×10^5	0.1749	0.3307	0.3079	0.2061	0.7959	0.2061
13	4.9207×10^5	0.1801	0.3607	0.2245	0.1351	0.8684	0.1351
14	5.2537×10^5	0.1850	0.3910	0.1357	0.0672	0.9414	0.0672
15	5.7623×10^5	0.1899	0.4153	0.0000	0.0000	1.0000	0.0000

The optimal values of the control variables among 18 generated load-wind scenarios for the BCS are represented in Table 17. As it can be observed from Table 17, the VAR sources gain the value of zero in most of the scenarios for the BCS. Hence, the fixed installation cost is calculated for the VAR sources that gain the value of zero.

In order to investigate the impact of VAR planning in bus voltage magnitude over the different load-wind scenarios, the voltage profile of load buses is plotted for each load-wind scenario in Figure 4. This Figure shows that the voltage magnitude of the load buses is limited to the range of 0.95 p.u. and 1.05 p.u. for all scenarios. As a result, the voltage magnitude of the load buses is regulated within the predefined limits.

Table 17. The optimal values of the control variables for probabilistic multi-objective RPP considering load demand and wind power generation uncertainties.

Control Variable	V_{g1} (p.u.)	V_{g2} (p.u.)	V_{g5} (p.u.)	V_{g8} (p.u.)	V_{g11} (p.u.)	V_{g13} (p.u.)	t_{6-9} (p.u.)	t_{6-10} (p.u.)	t_{4-12} (p.u.)	t_{28-27} (p.u.)	Q_{c24} (MVAR)	Q_{c25} (MVAR)	Q_{c26} (MVAR)	Q_{c29} (MVAR)	Q_{c30} (MVAR)
S_1	1.0622	1.0530	1.0766	1.0534	1.1000	1.0868	1.0271	0.9000	1.0103	0.9522	0.0000	0.0000	0.0000	0.0000	0.0000
S_2	1.0618	1.0528	1.0759	1.0539	1.1000	1.0854	1.0288	0.9000	1.0072	0.9563	0.0000	0.0000	0.0000	0.0000	1.1342
S_3	1.0610	1.0525	1.0754	1.0545	1.1000	1.0849	1.0311	0.9000	1.0047	0.9555	0.0000	0.0000	0.0000	0.0000	1.1557
S_4	1.0602	1.0521	1.0751	1.0545	1.1000	1.0846	1.0333	0.9000	1.0026	0.9546	0.0000	0.0000	0.0000	0.0000	1.1916
S_5	1.0594	1.0518	1.0747	1.0547	1.1000	1.0846	1.0351	0.9000	1.0007	0.9541	0.0000	0.0000	0.0000	0.0000	1.3510
S_6	1.0594	1.0517	1.0751	1.0545	1.1000	1.0856	1.0353	0.9000	1.0008	0.9532	0.0000	0.0000	0.0000	0.0000	1.4686
S_7	1.0631	1.0532	1.0775	1.0529	1.1000	1.0884	1.0257	0.9000	1.0114	0.9569	0.0000	0.0000	0.0000	0.0000	1.6385
S_8	1.0627	1.0530	1.0772	1.0530	1.1000	1.0881	1.0270	0.9000	1.0100	0.9558	0.0000	0.0000	0.0000	0.0000	1.4814
S_9	1.0619	1.0526	1.0768	1.0533	1.1000	1.0874	1.0294	0.9000	1.0073	0.9551	0.0000	0.0000	0.0000	0.0000	1.5306
S_{10}	1.0611	1.0523	1.0764	1.0536	1.1000	1.0870	1.0315	0.9000	1.0050	0.9540	0.0000	0.0000	0.0000	0.0000	1.5171
S_{11}	1.0603	1.0519	1.0760	1.0538	1.1000	1.0869	1.0335	0.9000	1.0029	0.9534	0.0000	0.0000	0.0000	0.0000	1.6548
S_{12}	1.0599	1.0518	1.0758	1.0540	1.1000	1.0868	1.0344	0.9000	1.0018	0.9535	0.0000	0.0000	0.0000	0.0000	1.8457
S_{13}	1.0639	1.0533	1.0787	1.0521	1.1000	1.0910	1.0240	0.9000	1.0141	0.9575	0.0000	0.0000	0.0000	0.0000	2.2814
S_{14}	1.0635	1.0531	1.0786	1.0521	1.1000	1.0908	1.0253	0.9000	1.0130	0.9555	0.0000	0.0000	0.0000	0.0000	1.8830
S_{15}	1.0628	1.0528	1.0781	1.0524	1.1000	1.0901	1.0277	0.9000	1.0102	0.9544	0.0000	0.0000	0.0000	0.0000	1.8246
S_{16}	1.0620	1.0524	1.0775	1.0527	1.1000	1.0895	1.0299	0.9000	1.0076	0.9544	0.0000	0.0000	0.0000	0.0000	2.0862
S_{17}	1.0610	1.0518	1.0764	1.0527	1.1000	1.0905	1.0316	0.9000	1.0075	0.9524	0.0000	0.0000	0.0000	0.0000	1.9060
S_{18}	1.0606	1.0515	1.0758	1.0528	1.1000	1.0910	1.0324	0.9000	1.0074	0.9518	0.0000	0.0000	0.0000	0.0000	1.9321
Expected Value	1.0611	1.0523	1.0764	1.0536	1.1000	1.0874	1.0314	0.9000	1.0053	0.9543	0.0000	0.0000	0.0000	0.0000	1.6427

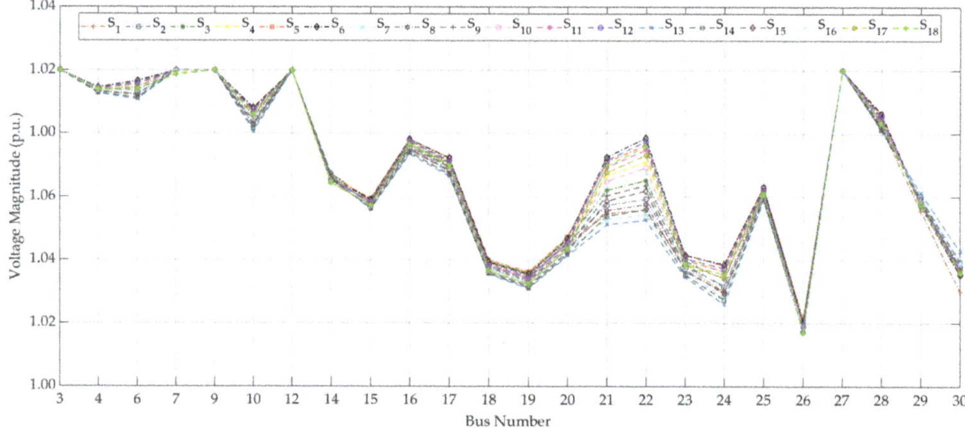

Figure 4. Voltage profile of the load buses for probabilistic multi-objective RPP considering the wind power generation uncertainty.

5.1.6. Case F: Probabilistic Multi-Objective RPP Considering Load Demand and Wind Power Generation Uncertainties Incorporating Reactive Power from Wind Farms

In order to evaluate the impact of generated reactive power by wind farms on RPP, the reactive power of wind farms is taken into account during the planning process. Assuming the constant PF operation for the proposed wind farms, the generated reactive power is calculated for each scenario based on Table 18. The value of PF for the proposed wind farms is taken to be 0.98. It should be noted that the generated reactive power by the proposed wind farms is calculated using Equation (47), as follows:

$$Q_{W_{i,S}} = P_{W_{i,S}} \tan\left(\cos^{-1}(PF)\right) \tag{47}$$

where $Q_{W_{i,S}}$ and $P_{W_{i,S}}$ indicate the generated reactive and active power by the proposed wind farms, respectively.

Table 18. The characteristics of generated reactive power by the proposed wind farm for different scenarios.

Scenario	Wind Power Generation (MVAR)
S_1	0.0000
S_2	1.0702
S_3	3.0645
S_4	5.0709
S_5	7.0824
S_6	8.1223
S_7	0.0000
S_8	1.0702
S_9	3.0645
S_{10}	5.0709
S_{11}	7.0824
S_{12}	8.1223
S_{13}	0.0000
S_{14}	1.0702
S_{15}	3.0645
S_{16}	5.0709
S_{17}	7.0824
S_{18}	8.1223

Similar to the former case studies, after performing the probabilistic multi-objective RPP considering the reactive power injection by the proposed wind farms, 15 Pareto optimal solutions are obtained, as shown in Table 19.

Table 19. Obtained Pareto optimal solutions for the probabilistic multi-objective RPP considering load demand and wind power generation uncertainties incorporating the generated reactive power by the proposed wind farms.

x	$f_1(\$)$	f_2	f_3	$\hat{F}_1$	$\hat{F}_2$	$\hat{F}_3$	$\min(\hat{F}_1,\hat{F}_2,\hat{F}_3)$
1	2.0015×10^5	0.1164	0.0010	1.0000	1.0000	0.0000	0.0000
2	2.1125×10^5	0.1215	0.0308	0.9706	0.9279	0.0719	0.0719
3	2.2251×10^5	0.1267	0.0608	0.9407	0.8559	0.1443	0.1443
4	2.4128×10^5	0.1319	0.0908	0.8910	0.7838	0.2167	0.2167
5	2.6295×10^5	0.1370	0.1208	0.8336	0.7118	0.2892	0.2892
6	2.8786×10^5	0.1422	0.1508	0.7676	0.6397	0.3616	0.3616
7	3.1517×10^5	0.1474	0.1808	0.6953	0.5677	0.4340	0.4340
8	3.4345×10^5	0.1525	0.2108	0.6203	0.4959	0.5064	0.4959
9	3.7331×10^5	0.1577	0.2408	0.5412	0.4240	0.5789	0.4240
10	4.0216×10^5	0.1629	0.2708	0.4648	0.3521	0.6513	0.3521
11	4.3098×10^5	0.1680	0.3008	0.3884	0.2802	0.7237	0.2802
12	4.6004×10^5	0.1732	0.3308	0.3115	0.2083	0.7961	0.2083
13	4.9099×10^5	0.1782	0.3609	0.2295	0.1377	0.8685	0.1377
14	5.2389×10^5	0.1831	0.3909	0.1423	0.0696	0.9410	0.0696
15	5.7760×10^5	0.1881	0.4153	0.0000	0.0000	1.0000	0.0000

From Table 19, it is clear that for the BCS, all the objectives are slightly improved towards Case E. Although this enhancement does not seem to be significant, it shows the penetration of generated reactive power by the proposed wind farms on planning studies. Moreover, the related active power loss reaches 8.4807 MW, which shows a reduction with respect to Case E. The optimal values of the control variables among 18 generated load-wind scenarios incorporating the generated reactive power by the proposed wind farms for the BCS are represented in Table 20.

Table 20. The optimal values of the control variables for probabilistic multi-objective RPP considering load demand and wind power generation uncertainties incorporating the generated reactive power by the proposed wind farms.

Control Variable	V_{g1} (p.u.)	V_{g2} (p.u.)	V_{g5} (p.u.)	V_{g8} (p.u.)	V_{g11} (p.u.)	V_{g13} (p.u.)	t_{6-9} (p.u.)	t_{6-10} (p.u.)	t_{4-12} (p.u.)	t_{28-27} (p.u.)	Q_{c24} (MVAR)	Q_{c25} (MVAR)	Q_{c26} (MVAR)	Q_{c29} (MVAR)	Q_{c30} (MVAR)
S_1	1.0622	1.0530	1.0766	1.0534	1.1000	1.0868	1.0271	0.9000	1.0103	0.9522	0.0000	0.0000	0.0000	0.0000	0.0000
S_2	1.0618	1.0528	1.0757	1.0541	1.1000	1.0845	1.0301	0.9000	1.0063	0.9568	0.0000	0.0000	0.0000	0.0000	1.0974
S_3	1.0609	1.0524	1.0748	1.0547	1.1000	1.0821	1.0350	0.9000	1.0020	0.9570	0.0000	0.0000	0.0000	0.0000	1.0878
S_4	1.0601	1.0521	1.0739	1.0553	1.1000	1.0800	1.0397	0.9000	0.9980	0.9572	0.0000	0.0000	0.0000	0.0000	1.1175
S_5	1.0594	1.0517	1.0732	1.0559	1.1000	1.0782	1.0441	0.9000	0.9945	0.9576	0.0000	0.0000	0.0000	0.0000	1.2400
S_6	1.0593	1.0516	1.0732	1.0558	1.1000	1.0782	1.0458	0.9000	0.9936	0.9576	0.0000	0.0000	0.0000	0.0000	1.4022
S_7	1.0631	1.0531	1.0773	1.0530	1.1000	1.0881	1.0258	0.9000	1.0109	0.9584	0.0000	0.0000	0.0000	0.0000	2.0174
S_8	1.0627	1.0530	1.0770	1.0532	1.1000	1.0871	1.0284	0.9000	1.0090	0.9563	0.0000	0.0000	0.0000	0.0000	1.4457
S_9	1.0619	1.0526	1.0761	1.0538	1.1000	1.0846	1.0333	0.9000	1.0045	0.9565	0.0000	0.0000	0.0000	0.0000	1.4365
S_{10}	1.0611	1.0522	1.0753	1.0544	1.1000	1.0824	1.0380	0.9000	1.0005	0.9565	0.0000	0.0000	0.0000	0.0000	1.3856
S_{11}	1.0603	1.0519	1.0744	1.0550	1.1000	1.0805	1.0425	0.9000	0.9967	0.9569	0.0000	0.0000	0.0000	0.0000	1.4942
S_{12}	1.0598	1.0517	1.0740	1.0553	1.1000	1.0795	1.0447	0.9000	0.9948	0.9572	0.0000	0.0000	0.0000	0.0000	1.5764
S_{13}	1.0639	1.0533	1.0787	1.0521	1.1000	1.0910	1.0240	0.9000	1.0141	0.9574	0.0000	0.0000	0.0000	0.0000	2.2599
S_{14}	1.0635	1.0531	1.0783	1.0523	1.1000	1.0898	1.0266	0.9000	1.0120	0.9559	0.0000	0.0000	0.0000	0.0000	1.8262
S_{15}	1.0628	1.0527	1.0774	1.0529	1.1000	1.0872	1.0315	0.9000	1.0073	0.9559	0.0000	0.0000	0.0000	0.0000	1.7669
S_{16}	1.0620	1.0524	1.0766	1.0535	1.1000	1.0849	1.0363	0.9000	1.0030	0.9561	0.0000	0.0000	0.0000	0.0000	1.7580
S_{17}	1.0612	1.0520	1.0758	1.0541	1.1000	1.0828	1.0408	0.9000	0.9990	0.9564	0.0000	0.0000	0.0000	0.0000	1.8356
S_{18}	1.0608	1.0518	1.0754	1.0544	1.1000	1.0819	1.0431	0.9000	0.9971	0.9566	0.0000	0.0000	0.0000	0.0000	1.9071
Expected Value	1.0610	1.0522	1.0753	1.0544	1.1000	1.0826	1.0380	0.9000	1.0006	0.9569	0.0000	0.0000	0.0000	0.0000	1.5296

It can be inferred from Table 3 that in almost all scenarios, the VAR sources gain the value of zero, except for the installed VAR compensator at bus 30. In addition, it is revealed that the expected value of the required VAR compensator device at bus 30 reduces while the generated reactive power of the hypothetical wind farms is taken into account. As a result, wind farms have the capability to participate in VAR planning. This leads to a reduction in the size and amount of VAR sources. Therefore, practical power systems show less desire to install new VAR support while numerous large-scale wind farms with sufficient generated reactive power are available.

5.2. Discussions

Table 21 compares the performance of Case E with other cases. As it can be observed, compared with Case B, the performance of Case E in terms of obtaining better values for f_1, f_2, and P_{loss} is improved. Case E also shows better performance rather than Case C. All objectives are improved considerably compared with Case C. Note that, due to the presence of wind farms, f_2 is enhanced, while improving on the loadability index. In addition, Case E is compared with Case D, and it can be observed that all objectives are improved. f_1 is improved significantly. However, f_2, f_3, and P_{loss} are not enhanced considerably. This is due to the fact that wind power generation uncertainty has a great impact on all objectives, which in both Case E and Case D are considered. While, the performance of Case F is better than Case E considering f_1, f_2, and P_{loss}, and is slightly more than f_3.

Table 21. Comparison of different cases.

Case	f_1 (\$)	f_2	f_3	P_{loss} (MW)
B	4.9361×10^6	0.1533	0.1653	9.3494
C	2.2510×10^6	0.1606	0.1688	9.5049
D	7.5741×10^5	0.1570	0.2100	8.5777
E	3.4683×10^5	0.1539	0.2107	8.5575
F	3.4345×10^5	0.1525	0.2108	8.4807

6. Conclusions

A multi-objective RPP in power systems considering load demand and wind power generation uncertainties to minimize reactive power investment cost, reduce active power losses, improve voltage stability level, and enhance loadability factor is presented in this paper. The ε-Constraint method is used to solve the probabilistic multi-objective RPP. For this purpose, using the L-index, the VAR compensation buses are found at the first stage. Then, to distinguish the exact difference between the deterministic and probabilistic VAR planning studies, five different cases are investigated. In order to test the efficiency of the proposed method, the IEEE 30-bus test system is implemented in GAMS software under five various conditions. The simulation results show that the proposed probabilistic multi-objective RPP considering load demand and wind power generation uncertainties is effective in reducing the VAR installation cost, improving the voltage stability of the system, and enhancing the loadability, simultaneously.

Author Contributions: All authors participated equally in conceptualization, methodology, implementation, and writing—review and editing the paper. Also, All authors have read and agreed to the published version of the manuscript.

Funding: This research received no external funding.

Conflicts of Interest: The authors declare no conflict of interest.

Appendix A

Table A1. Parameters of the wind farm.

Parameter	Value
α	2
β	10
v_{in}^c	3 m/s
v_{rated}	10.28 m/s
v_{out}^c	25 m/s
P_w^r	40 MW

References

1. Zhang, W.; Li, F.; Tolbert, L.M. Review of Reactive Power Planning: Objectives, Constraints, and Algorithms. *IEEE Trans. Power Syst.* **2007**, *22*, 2177–2186. [CrossRef]
2. Seifi, H.; Sepasian, M.S. *Electric Power System Planning: Issues, Algorithms and Solutions*; Springer: Berlin/Heidelberg, Germany, 2011.
3. Roselyn, J.P.; Devaraj, D.; Dash, S.S. Multi Objective Differential Evolution Approach for Voltage Stability Constrained Reactive Power Planning Problem. *Int. J. Electr. Power Energy Syst.* **2014**, *59*, 155–165. [CrossRef]
4. Miveh, M.R.; Rahmat, M.F.; Ghadimi, A.A.; Mustafa, M.W. Control Techniques for Three-Phase Four-Leg Voltage Source Inverters in Autonomous Microgrids: A Review. *Renew. Sustain. Energy Rev.* **2016**, *54*, 1592–1610. [CrossRef]
5. Miveh, M.R.; Rahmat, M.F.; Mustafa, M.W.; Ghadimi, A.A.; Rezvani, A. An Improved Control Strategy for a Four-Leg Grid-Forming Power Converter under Unbalanced Load Conditions. *Adv. Power Electron.* **2016**, *2016*. [CrossRef]
6. Mohammadi, F.; Nazri, G.-A.; Saif, M. A Fast Fault Detection and Identification Approach in Power Distribution Systems. In Proceedings of the 5th International Conference on Power Generation Systems and Renewable Energy Technologies (PGSRET), Istanbul, Turkey, 26–27 August 2019.
7. Han, T.; Chen, Y.; Ma, J.; Zhao, Y.; Chi, Y.-Y. Surrogate Modeling-Based Multi-Objective Dynamic VAR Planning Considering Short-Term Voltage Stability and Transient Stability. *IEEE Trans. Power Syst.* **2018**, *33*, 622–633. [CrossRef]
8. Raj, S.; Bhattacharyya, B. Optimal Placement of TCSC and SVC for Reactive Power Planning Using Whale Optimization Algorithm. *Swarm Evol. Comput.* **2018**, *40*, 131–143. [CrossRef]
9. Bhattacharyya, B.; Raj, S. Swarm Intelligence Based Algorithms for Reactive Power Planning with Flexible AC Transmission System Devices. *Int. J. Electr. Power Energy Syst.* **2016**, *78*, 158–164. [CrossRef]
10. Alonso, M.; Amaris, H.; Alvarez-Ortega, C. A Multiobjective Approach for Reactive Power Planning in Networks with Wind Power Generation. *Renew. Energy* **2012**, *37*, 180–191. [CrossRef]
11. Amaris, H.; Alonso, M. Coordinated Reactive Power Management in Power Networks with Wind Turbines and FACTS Devices. *Energy Convers. Manag.* **2011**, *52*, 2575–2586. [CrossRef]
12. Fang, X.; Li, F.; Wei, Y.; Azim, R.; Xu, Y. Reactive Power Planning Under High Penetration of Wind Energy Using Benders Decomposition. *IET Gener. Transm. Distrib.* **2015**, *9*, 1835–1844. [CrossRef]
13. Niu, M.; Xu, Z. Reactive Power Planning for Transmission Grids with Wind Power Penetration. In Proceedings of the IEEE PES Innovative Smart Grid Technologies, Tianjin, China, 21–24 May 2012.
14. López, J.C.; Contreras, J.; Muñoz, J.I.; Mantovani, J. A Multi-Stage Stochastic Non-Linear Model for Reactive Power Planning Under Contingencies. *IEEE Trans. Power Syst.* **2013**, *28*, 1503–1514. [CrossRef]
15. López, J.; Pozo, D.; Contreras, J.; Mantovani, J.R.S. A Multiobjective Minimax Regret Robust VAR Planning Model. *IEEE Trans. Power Syst.* **2017**, *32*, 1761–1771. [CrossRef]
16. Yang, N.; Yu, C.; Wen, F.; Chung, C. An Investigation of Reactive Power Planning Based on Chance Constrained Programming. *Int. J. Electr. Power Energy Syst.* **2007**, *29*, 650–656. [CrossRef]
17. López, J.C.; Mantovani, J.S.; Sanz, J.C.; Muñoz, J.I. Optimal Reactive Power Planning Using Two-Stage Stochastic Chance-Constrained Programming. In Proceedings of the IEEE Grenoble Conference, Grenoble, France, 16–20 June 2013.

18. Aien, M.; Hajebrahimi, A.; Fotuhi-Firuzabad, M. A Comprehensive Review on Uncertainty Modeling Techniques in Power System Studies. *Renew. Sustain. Energy Rev.* **2016**, *57*, 1077–1089. [CrossRef]

19. Tabatabaei, N.M.; Aghbolaghi, A.J.; Bizon, N.; Blaabjerg, F. *Reactive Power Control in AC Power Systems: Fundamentals and Current Issues*; Springer International Publishing: Berlin/Heidelberg, Germany, 2017.

20. Aien, M.; Fotuhi-Firuzabad, M.; Rashidinejad, M. Probabilistic Optimal Power Flow in Correlated Hybrid Wind–Photovoltaic Power Systems. *IEEE Trans. Smart Grid* **2014**, *5*, 130–138. [CrossRef]

21. Aien, M.; Khajeh, M.G.; Rashidinejad, M.; Fotuhi-Firuzabad, M. Probabilistic Power Flow of Correlated Hybrid Wind-Photovoltaic Power Systems. *IET Renew. Power Gener.* **2014**, *8*, 649–658. [CrossRef]

22. Growe-Kuska, N.; Heitsch, H.; Romisch, W. Scenario Reduction and Scenario Tree Construction for Power Management Problems. In Proceedings of the 2003 IEEE Bologna Power Tech Conference Proceedings, Bologna, Italy, 23–26 June 2003.

23. Vagropoulos, S.I.; Kardakos, E.G.; Simoglou, C.K.; Bakirtzis, A.G.; Catalao, J.P. ANN-Based Scenario Generation Methodology for Stochastic Variables of Electric Power Systems. *Electr. Power Syst. Res.* **2016**, *134*, 9–18. [CrossRef]

24. Mohseni-Bonab, S.M.; Rabiee, A. Optimal Reactive Power Dispatch: A Review, and A New Stochastic Voltage Stability Constrained Multi-Objective Model at the Presence of Uncertain Wind Power Generation. *IET Gener. Transm. Distrib.* **2017**, *11*, 815–829. [CrossRef]

25. Mohseni-Bonab, S.M.; Rabiee, A.; Mohammadi-Ivatloo, B. Voltage Stability Constrained Multi-Objective Optimal Reactive Power Dispatch under Load and Wind Power Uncertainties: A Stochastic Approach. *Renew. Energy* **2016**, *85*, 598–609. [CrossRef]

26. López, J.C.; Muñoz, J.I.; Contreras, J.; Mantovani, J. Optimal Reactive Power Dispatch Using Stochastic Chance-Constrained Programming. In Proceedings of the 2012 Sixth IEEE/PES Transmission and Distribution: Latin America Conference and Exposition (T&D-LA), Montevideo, Uruguay, 3–5 September 2012.

27. Mohammadi, F.; Zheng, C. Stability Analysis of Electric Power System. In Proceedings of the 4th National Conference on Technology in Electrical and Computer Engineering, Tehran, Iran, 27 December 2018.

28. Mohammadi, F.; Nazri, G.-A.; Saif, M. A Bidirectional Power Charging Control Strategy for Plug-in Hybrid Electric Vehicles. *Sustainability* **2019**, *11*, 4317. [CrossRef]

29. Mohammadi, F.; Nazri, G.-A.; Saif, M. An Improved Mixed AC/DC Power Flow Algorithm in Hybrid AC/DC Grids with MT-HVDC Systems. *Appl. Sci.* **2020**, *10*, 297. [CrossRef]

30. Ma, J.; Lai, L.L. Evolutionary Programming Approach to Reactive Power Planning. *IEE Proc. Gener. Transm. Distrib.* **1996**, *143*, 365–370. [CrossRef]

31. Kessel, P.; Glavitsch, H. Estimating the Voltage Stability of a Power System. *IEEE Trans. Power Deliv.* **1986**, *1*, 346–354. [CrossRef]

32. Singh, J.; Singh, S.; Srivastava, S. Placement of FACTS Controllers for Enhancing Power System Loadability. In Proceedings of the 2006 IEEE Power India Conference, New Delhi, India, 10–12 April 2006.

33. Kalyanmoy, D. *Multi-Objective Optimization using Evolutionary Algorithms*; John Wiley and Sons: Hoboken, NJ, USA, 2001.

34. Cohon, J.L. *Multiobjective Programming and Planning*; Courier Corporation: North Chelmsford, MA, USA, 2004.

35. Wang, X.; Gong, Y.; Jiang, C. Regional Carbon Emission Management Based on Probabilistic Power Flow with Correlated Stochastic Variables. *IEEE Trans. Power Syst.* **2015**, *30*, 1094–1103. [CrossRef]

36. Rosenthal, R.E. *User's Guide, Tutorial, GAMS*; Development Corporation: Washington, DC, USA, 2010.

37. Andrei, N. *Nonlinear Optimization Applications Using the GAMS Technology*; Springer: Berlin/Heidelberg, Germany, 2013.

38. Soroudi, A. *Power System Optimization Modeling in GAMS*; Springer: Berlin/Heidelberg, Germany, 2017.

39. Ferris, M.C.; Jain, R.; Dirkse, S. GDXMRW: Interfacing GAMS and MATLAB. 2010. Available online: http://www.gams.com/dd/docs/tools/gdxmrw (accessed on 15 March 2020).

40. Drud, A. *CONOPT Solver Manual*; ARKI Consulting and Development: Bagsværd, Denmark, 1996.

41. Lee, K.; Park, Y.; Ortiz, J. A United Approach to Optimal Real and Reactive Power Dispatch. *IEEE Trans. Power Appar. Syst.* **1985**, *PAS-104*, 1147–1153. [CrossRef]

42. Christie, R. *Power Systems Test Case Archive*; UW Power Systems Test Case Archive: Seattle, WA, USA, 1993.

43. Nguyen, T.T.; Mohammadi, F. Optimal Placement of TCSC for Congestion Management and Power Loss Reduction Using Multi-Objective Genetic Algorithm. *Sustainability* **2020**, *12*, 2813. [CrossRef]

44. Mahmoudabadi, A.; Rashidinejad, M. An Application of Hybrid Heuristic Method to Solve Concurrent Transmission Network Expansion and Reactive Power Planning. *Int. J. Electr. Power Energy Syst.* **2013**, *45*, 71–77. [CrossRef]
45. Mishra, C.; Singh, S.P.; Rokadia, J. Optimal Power Flow in the Presence of Wind Power Using Modified Cuckoo Search. *IET Gener. Transm. Distrib.* **2015**, *9*, 615–626. [CrossRef]

Article

Effect of Ionic Conductors on the Suppression of PTC and Carrier Emission of Semiconductive Composites

Yingchao Cui [1], Hongxia Yin [1], Zhaoliang Xing [2], Xiangjin Guo [1], Shiyi Zhao [1], Yanhui Wei [1], Guochang Li [1], Meng Xin [1], Chuncheng Hao [1,2,*] and Qingquan Lei [1]

[1] Institute of Advanced Electrical Materials, Qingdao University of Science and Technology, Qingdao 266042, China; cc18753214515@163.com (Y.C.); 13455018936@163.com (H.Y.); Guoaiwei525@163.com (X.G.); zsy19941103@126.com (S.Z.); weiyhui@126.com (Y.W.); Lgc@qust.edu.cn (G.L.); xinmeng_7591@126.com (M.X.); leiqingquan@qust.edu.cn (Q.L.)

[2] State Key Laboratory of Advanced Power Transmission Technology (Global Energy Interconnection Research Institute Co., Ltd.), Beijing 102209, China; xingzhaoliang007@163.com

* Correspondence: clx@qust.edu.cn; Tel.: +86-187-0532-1299

Received: 24 February 2020; Accepted: 18 April 2020; Published: 23 April 2020

Abstract: The positive temperature coefficient (PTC) effect of the semiconductive layers of high-voltage direct current (HVDC) cables is a key factor limiting its usage when the temperature exceeds 70 °C. The conductivity of the ionic conductor increases with the increase in temperature. Based on the characteristics of the ionic conductor, the PTC effect of the composite can be weakened by doping the ionic conductor into the semiconductive materials. Thus, in this paper, the PCT effects of electrical resistivity in perovskite $La_{0.6}Sr_{0.4}CoO_3$ (LSC) particle-dispersed semiconductive composites are discussed based on experimental results from scanning electron microscopy (SEM), transmission electron microscopy (TEM) and a semiconductive resistance test device. Semiconductive composites with different LSC contents of 0.5 wt%, 1 wt%, 3 wt%, and 5 wt% were prepared by hot pressing crosslinking. The results show that the PTC effect is weakened due to the addition of LSC. At the same time, the injection of space charge in the insulating sample is characterized by the pulsed electroacoustic method (PEA) and the thermally stimulated current method (TSC), and the results show that when the content of LSC is 1 wt%, the injection of space charge in the insulating layer can be significantly reduced.

Keywords: $La_{0.6}Sr_{0.4}CoO_3$; semiconductive layer; PTC effect; space charge; HVDC transmission

---

1. Introduction

High-voltage direct current (HVDC) transmission plays a significant role in the power system [1–5]. In particular, HVDC cable transmission is feasible over long distances and large capacities due to the absence of reactive power and low transmission losses [6,7]. Typical medium and high-voltage power polyethylene (PE)cable cross-sectional constructions include: (1) conductors, (2) conductor shield, (3) insulation, (4) insulation shield, (5) metal shield, and (6) enclosure material [8]. In the construction of high-voltage power cables, the semiconductive layer can suppress the injection of carriers from the metal electrode into the insulating layer and can effectively prevent local electric field distortion between the conductor and the insulating layer.

However, the electrical resistance of the semiconductive layer can suddenly increase to 90 °C, which causes the cable to heat up and leads the interface to partially melt. This phenomenon is called the positive temperature coefficient (PTC) effect [9]. The PTC effect of semiconductive composites is usually weakened by increasing the content of carbon black (CB) or by using high-structure carbon black [10,11]. However, the amount of CB added to the semiconductive shielding layer affects its processing and mechanical properties. The conductivity of the ionic conductor increases with increasing temperature. In this work, the influence of the (CB—La$_{0.6}$Sr$_{0.4}$CoO$_3$ (LSC)) co-filled on the electrical properties of semiconductive composites was studied, in which LSC was used as a second filler to suppress the PTC effect. The perovskite oxide LaCoO$_3$ has been widely used because of its high ionic and electrical conductivity. The ideal perovskite structure is shown in Figure 1 [12–14]. When La in LaCoO$_3$ is partly replaced by the Sr, the lattice spacing becomes larger and the oxygen vacancies in the crystal increase [15]. Oxygen vacancies and lattice defects of LSC can provide more conductive channels for electrons, which facilitates electron migration when Sr-doped LaCoO$_3$ is added to a semiconductive composite material.

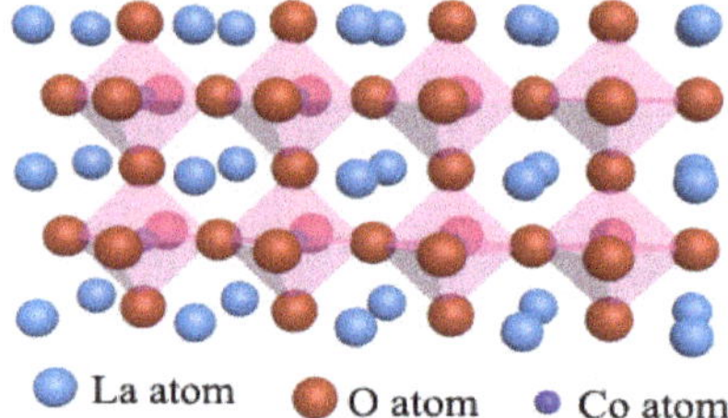

Figure 1. Structure of LaCoO$_3$.

On the other hand, space charge accumulation of the insulating layer is another key factor affecting the stable operation of the cable. Insulating layers tend to accumulate space charge, which causes distortion of the electric field [16,17]. Eventually, the insulating layer is easily aging or mangled [18–20]. Until recently, most studies have been limited to the insulation nano-doped polyethylene, which has been attracting more and more attention. It has been reported that polyethylene-doped inorganic nanoparticles such as MgO, ZnO and SiO$_2$ can significantly suppress the accumulation of space charge in the insulation [21–25]. Some researchers add SrFe$_{16}$O$_{19}$ to the semiconductive layer to reduce the injection of space charge in the insulating layer by using the Lorentz force of magnetic particles on the charge [26]. However, in this study, we suppressed the injection of space charge by using the LSC modified semiconductive layer. The injection of charges into the insulating layer is reduced, through the Coulomb effect between LSC particles in semiconductive materials and injected charges. This work provides a new idea for the development of semiconductive materials.

2. Materials and Methods

2.1. Materials

2.1.1. Preparation of LSC

The LSC was prepared by using the sol-gel method [27,28], mixing a stoichiometric amount of lanthanum nitrate hexahydrate La(NO$_3$)$_3$·6H$_2$O strontium nitrate Sr(NO$_3$)$_2$ with cobalt nitrate hexahydrate Co(NO$_3$)$_2$·6H$_2$O in deionized water under constant stirring to get a clear solution. Citric acid (CA) was then added into the solution (CA and total metal ion in a 7:5 molar ratio), in which as a ligand to form a complex compound with the metal ion. Then, the pH value of the solution was adjusted to 9–10 by dropwise addition of aqueous ammonium hydroxide. The solution was slowly evaporated in a water bath at 70 °C for 10 h and the gel obtained was at the temperature of 150 °C overnight. Finally, the obtained powder was calcined at 900 °C for 6 h to obtain LSC nanoparticles.

2.1.2. Ball Milling of LSC

The prepared LSC powder was ball milled in a planetary ball mill, where 10 g of LSC powder and zirconia balls (mass ratio of zirconia balls to LSC powder of 20:1) were added to a ball mill jar, and then 200 mL of ethanol were added. The speed during ball milling was 300 rpm. The samples of ball milling of 20 and 40 h, respectively, were obtained for observation by scanning electron microscopy. Finally, the samples were dried at 50 °C to obtain the LSC nanoparticles after ball milling [29].

2.1.3. Preparation of the Nanocomposite

The matrix polymer was prepared by mixing 25% carbon black (CB), 45% low-density polyethylene (LDPE), and 30% ethylene-vinyl acetate copolymer (EVA) with an open mill at °C. Then, the LSC was mixed with the above matrix polymer in different mass percentages, as shown in Table 1. At last, the above materials were shaped by hot pressing by a vulcanizer.

Table 1. Sample notation and composition.

Sample	1#	2#	3#	4#	5#
CB/LDPE/EVA matrix (wt%)	100	99.5	99	97	95
LSC (wt%)	0	0.5	1	3	5

2.2. Characterization

2.2.1. X-ray Diffraction (XRD)

The crystal structural analyses of LSC were determined by XRD measurements (Rigaku, D/max-2500/PC) from 20° to 90°.

2.2.2. Scanning Electron Microscopy (SEM)

The morphology of the nanoparticles was observed with a field emission SEM (FEI·Nova·Nano·SEM450) at a 5 kV accelerating voltage. Dispersion of the nanoparticles in the semiconductive composites was observed using SEM (JSM-6700F, JEOL, Tokyo, Japan). The nanocomposites were broken in liquid nitrogen and then fractured cross-sections were sprayed with gold to avoid the charge accumulation effect during observation.

2.2.3. Transmission Electron Microscopy (TEM)

In order to characterize the dispersion of the LSC in the matrix polymer, 50–100 nm thick ultra-thin sections were cut using a ultramicrotome and observed using a TEM (FEI·Tecnai·G2F30).

2.2.4. Resistivity Test

In the actual operation of the cable, the working temperature is greatly affected by the load. The resistance of the semiconductive layer will increase with the increase in the temperature, showing obvious PTC effect, which will lead to the increase in the interface thermal effect between the semiconducting layer and the insulating layer, and affect the service life of the cable. In this work, the resistivity of the semiconductive layer was measured by the DB-4 wire and the cable semiconductive rubber-resistance tester using the (DC) current-voltage method test principle. The samples, with length 110 mm, width 50 mm, and thickness of 1 mm, were obtained by hot pressing crosslinked. The sample is placed in a drying oven with programmable temperature control, and the resistivity of the sample is recorded at different temperatures. When the instrument is used to measure, the sample does not need surface treatment, and the operation is simple. The resistivity of the sample can be obtained directly without formula derivation and calculation, thus avoiding the error in the calculation process.

2.2.5. Pulsed Electroacoustic Measurement (PEA)

The distribution of space charge was tested by PEA. The LDPE insulating sample used for PEA testing had an average thickness of 300 μm and the semiconductive layer had a thickness of 500 μm. The experiment was carried out for 30 min at room temperature under a negative DC electric field of 10 and 40 kV/mm. The PEA test chart is shown in Figure 2.

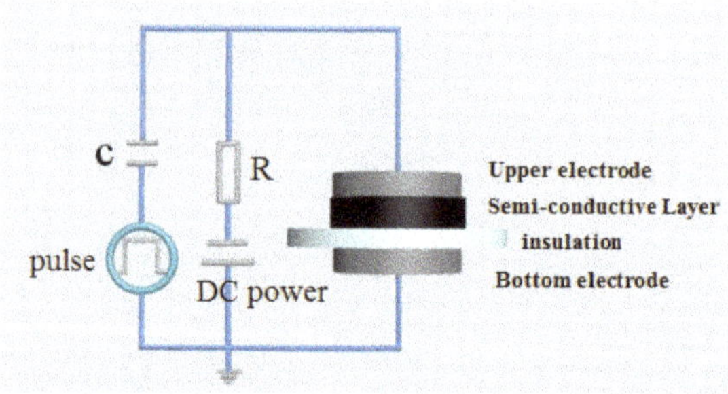

Figure 2. Illustration of the pulsed electroacoustic (PEA) measurement.

2.2.6. Thermally Stimulated Current (TSC)

The TSC method includes the thermal stimulation polarization current method (TSPC) and the thermal stimulation depolarization current method (TSDC). The TSDC method is more common in the measurement and characterization of traps in polymer insulation. Thus, the TSC method generally refers to the TSDC method. The TSDC method was used in this experiment. The thickness of the insulating sample and the semiconductive layer used for TSC was 300 and 500 μm, respectively. A negative DC field strength of 10, 30, and 40 kV/mm was applied to both ends of the LDPE for 30 min at room temperature when the semiconductive composites with different LSC contents were used as the semiconductive layer, and then, the sample was rapidly cooled. Next, the temperature was raised from 293 K at a heating rate of 5 K/min to 363 K to measure the value of the thermal stimulation current during the heating process. The schematic diagram of the TSDC test is shown in Figure 3.

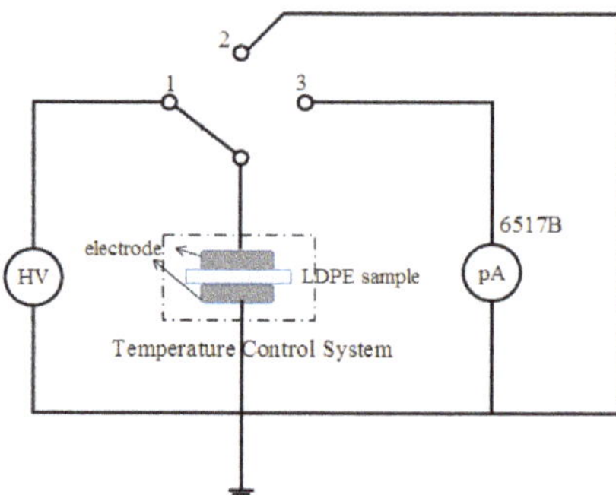

Figure 3. Schematic diagram of thermal stimulation current method.

3. Results

3.1. Structural Characterization of LSC

The XRD pattern of LSC is presented in Figure 4. It is observed that there are no impurity peaks from the XRD pattern, and the XRD diagram of LSC shows the characteristic of sharp peaks, indicating that the crystallization of LSC was excellent. The XRD pattern of LSC displays characteristic peaks at 2θ = (23.4°, 33.2°, 40.8°, 47.6°, 53.5°, 59.1°, 69.6°, 79.2°, 83.7°, and 88.2°,) which correspond to the planes of (012), (110), (202), (024), (122), (300), (220), (134), (042), and (404) simultaneously, consistent with the standard reference data (JCPDF:89-5719).

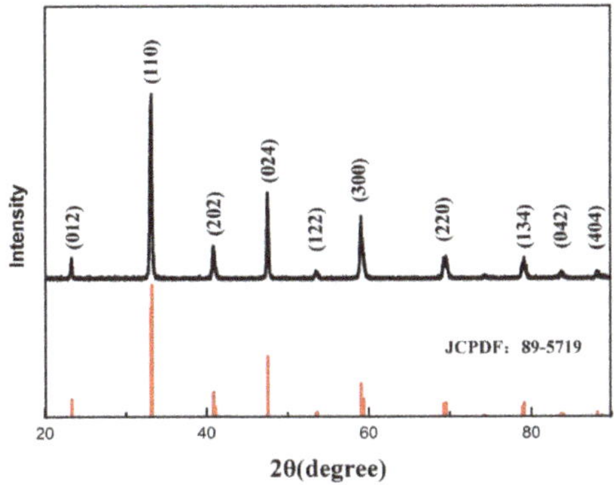

Figure 4. X-ray diffraction (XRD) patterns of LSC.

3.2. SEM of LSC

The SEM images of the LSC before ball milling and after ball milling for 10 and 20 h are shown in Figure 5. It can be seen from Figure 5 that the particle size of the nanoparticles decreases with increasing ball milling time. The LSC without ball milling is composed of particles with a particle size of about 300 nm. In Figure 5a, the grains of LSC powder are bonded together. After ball milling, the particles originally bonded together are dispersed. From Figure 5b, it can be seen that after ball milling for 10 h, the size of LSC particles is distributed around 400–700 nm. After ball milling for 20 h, the size of LSC particles is 300 nm. The particles bonded after ball milling are dispersed, and the particle size distribution is more uniform. This facilitates the preparation of a smooth semiconductive shielding layer.

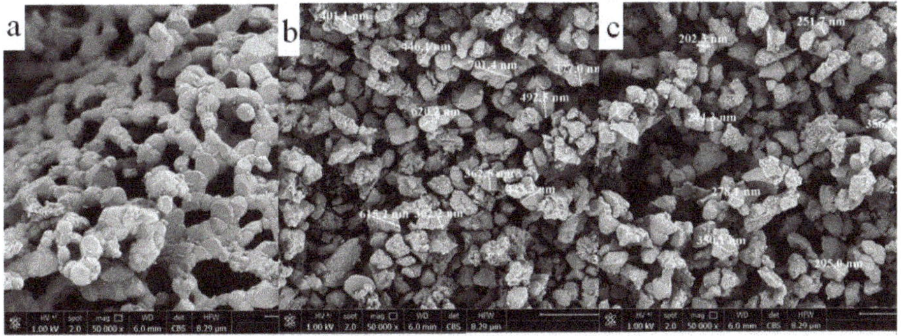

Figure 5. Scanning electron microscopy (SEM) images of LSC: (**a**) Ball milling (BM)-LSC (0 h); (**b**) BM-LSC (10 h); (**c**) BM-LSC (20 h).

3.3. SEM and TEM of the Semiconductive Shielding

The dispersion of nanoparticles in the matrix polymer can be observed by the SEM of Figure 6 and the TEM of Figure 7. Figure 6a–c shows the fracture surface SEM images of composite nanomaterials with an LSC content of 0%, 1%, and 5%, respectively. The white spots in Figure 6 are the LSC nanoparticles. It can be seen from Figure 6b that the nanoparticles are uniformly dispersed in the matrix and the white spots in Figure 6b,c increase as the LSC content increases. Figure 7 shows that, the contrast degree of the carbon black particles in the polymer matrix are light, and the black particles with deep contrast are LSC particles. It can be seen from the figure that the size of the black particles acts at several hundred nm, which matches the SEM image of the LSC particles in Figure 5. Figure 7 shows that carbon black particles fill the matrix polymer and form conductive channels. LSC particles are uniformly dispersed in the polymer matrix. However, the nanoparticles are prone to agglomeration when the LSC concentration is high.

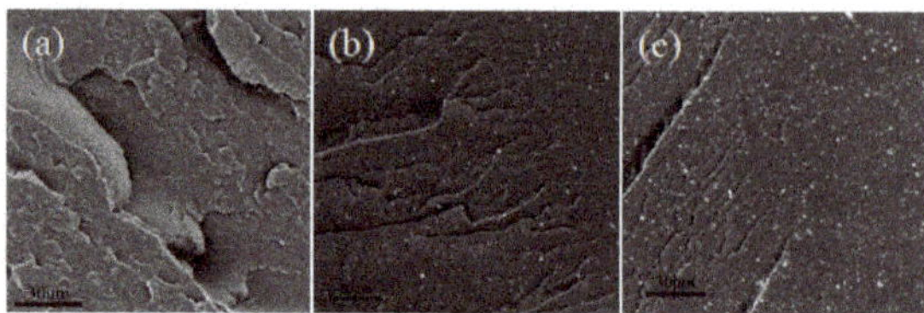

Figure 6. SEM of sections of non-semiconducting shielding materials with different LSC contents: (**a**) 0 wt% LSC; (**b**) 1 wt% LSC; (**c**) 5 wt% LSC.

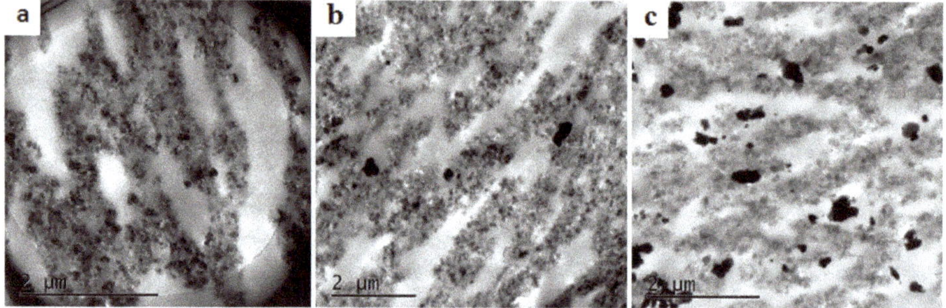

Figure 7. Transmission electron microscopy (TEM) micrographs of the semiconductive materials with different LSC contents: (**a**) 0 wt% LSC; (**b**) 0.5 wt% LSC; (**c**) 3 wt% LSC.

3.4. Electrical Properties of the Semiconductive Shielding

Figure 8 shows that curve of resistivity versus temperature for the semiconductive layer containing different mass fractions of LSC. Figure 9 shows the resistivity curve of semiconductive materials with different LSC contents at 383 K. It can be seen from Figure 9 that at 383 K, the resistivity of semiconducting shielding material without LSC doping is 798 $\rho/\Omega\cdot$cm and when the LSC doping amount is 1 wt%, the resistivity is 128.5 ρ/Ω cm, decreased by 83.9%. Some researchers have added $SrFe_{16}O_{19}$ to semiconductor shielding materials and tested their resistivity. The resistivity of semiconductive materials with $SrFe_{16}O_{19}$ doping of 1 wt% and 5 wt% is similar to that without $SrFe_{16}O_{19}$ doping. When the doping amount of $SrFe_{16}O_{19}$ is 30 wt%, the resistivity of semiconductive materials is more than 10^3 at 383 K [26]. We can see that the resistivity demonstrates a slow rising tendency with temperature before the temperature is below 343 K. Meanwhile, there is a huge transition in the resistivity value of the semiconductive composites without added LSC after the temperature exceeds 343 K. In other words, the semiconductive layer without added LSC possesses a significant PTC effect. Since the electrical conductivity of the LSC increases with increasing temperature, the semiconductive layer to which LSC is added still has good electrical conductivity at high temperatures. It can be seen from Figure 9 that at 383 K, the resistivity of semiconductive materials with 1 wt% LSC doping is greatly reduced compared with that without LSC. Therefore, the addition of LSC can improve the PTC effect of the semiconductive composites so that it still meets the resistivity requirements of the semiconductive layer at high temperatures. In particular, the semiconductive layer with a 1% LSC presents good electrical conductivity at high temperatures. This might be attributed to the distortion of the crystal structure of Sr-doped $LaCoO_3$, the lattice spacing becomes larger, and the amount of O vacancies increases, providing more conductive channels for carrier transport.

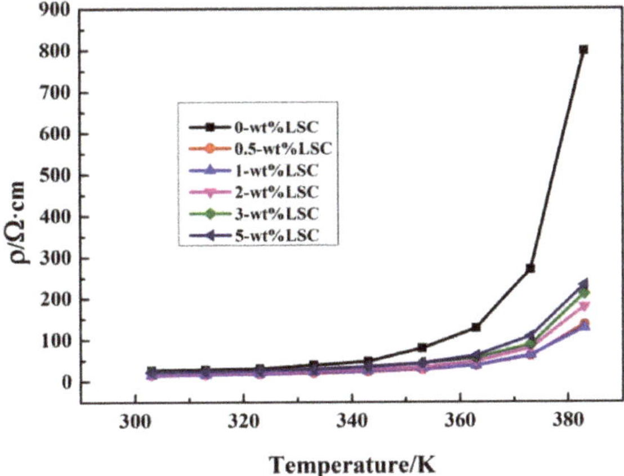

Figure 8. Resistivity of the semiconductive composites with different mass fractions of LSC as a function of temperature.

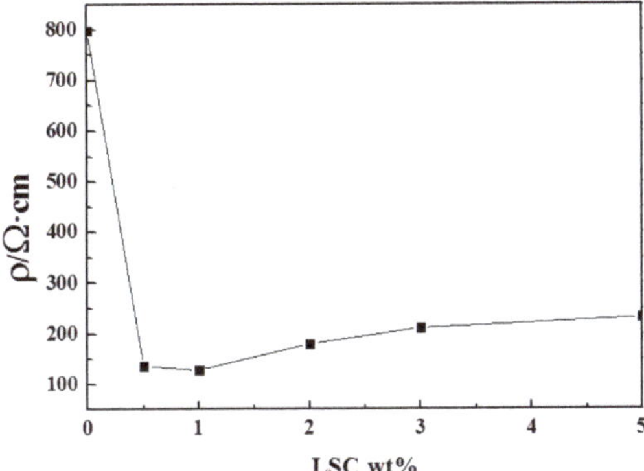

Figure 9. Resistivity curves of semiconductive materials with different LSC contents at 383 K.

The formula for calculating the strength of the polymer's positive temperature coefficient:

$$\alpha = \lg\frac{\rho_{v(\mathrm{max})}}{\rho_{v(\mathrm{min})}}$$

The calculated PTC strengths of semiconductive composites with LSC doping contents of 0%, 0.5 wt%, 1 wt%, 2 wt%, 3 wt% and 5 wt% were 1.47, 0.96, 0.91, 0.99, 0.95, 1.02. Compared with the PTC strength of the semiconductive materials without LSC, the PTC strength of the semiconductive materials with 1 wt% LSC content decreased by 38.1%, which indicated that the addition of LSC has a significant weakening effect on the PTC effect of the system, which is related to the increase in the conductivity of the ionic conductor with the increase in temperature. As the temperature increases, the number of carriers in the LSC increases, and the mobility of the carriers increases. Therefore, the PTC effect of the LSC/CB/LDPE/EVA composites is weaker. With the increase in LSC content, the PTC strength of nanocomposites decreases first and then increases. Because of the agglomeration of LSC in semiconductive materials, part of the carbon black conductive network in the composites is disconnected, thus, the PTC strength of semiconductive materials increases when LSC content is high.

3.5. Depolarization Current Properties

Figure 10 presents the depolarization current of the insulating layer when nanocomposites with different LSC contents were used as semiconductive layers. Figure 10a shows the depolarization current in LDPE at a 10 kV/mm DC field. It can be seen that the depolarization current increases first and then decreases with increasing temperature. The peak value of the current of all samples appeared at 330–340 K under 10 kV/mm DC field, which indicates that the trap levels are basically the same. At high loading levels, the peak value of the depolarization current of the LDPE increases as the LSC content in the semiconductive layer increases. Figure 10b,c shows the depolarization current of LDPE at 30 and 40 kV/mm. The depolarization current increases as the electric field increases, mainly because of the increased charge injection under a strong electric field. At the same time, the position of the peak moves toward the high temperature direction, mainly because the depth of charge injection increases as the electric field strength increases.

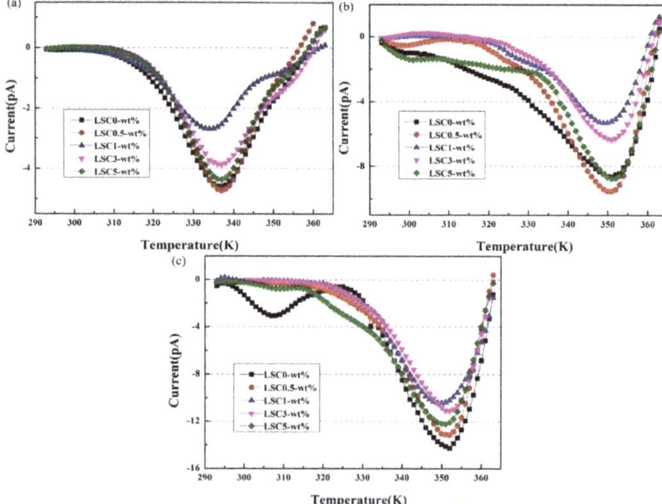

Figure 10. Thermal stimulation current of LDPE when different semiconductive layers are used as electrodes, the applied electric field was: (**a**) 10 kV/mm, (**b**) 30 kV/mm, (**c**) 40 kV/mm at room temperature.

The depolarization current peak that appears between 300 and 320 K in Figure 10c is due to the dipole polarization of small molecular chains and polar groups in LDPE.

The total trap charge can be calculated according to the TSDC curves. Figure 11 shows the amount of trap charge in LDPE when a composite with different LSC contents is used as a semiconductive layer. It can be concluded from Figure 11 that the effect of suppressing space charge injection when the composite material with an LSC content of 1% is used as the semiconductive layer is the most obvious. When the composites without LSC were used as the semiconductive layer, the charge amount in the insulating sample is 1.35×10^{-9}, 3.26×10^{-9}, and 4.26×10^{-9}, respectively, under 10, 30 and 40 kV/mm DC electric fields. For LSC content with 1 wt%, the charge of the insulating layer decreased to 0.75×10^{-9}, 1.34×10^{-9}, and 2.75×10^{-9}, respectively, decreasing by 44.4%, 58.9%, and 35.7%. When the LSC concentration in the semiconductive composites is high, the trap charge amount in the LDPE increases. The reason might be that the agglomeration of nanoparticles causes the surface roughness of the nanocomposite to increase, resulting in electric field distortion.

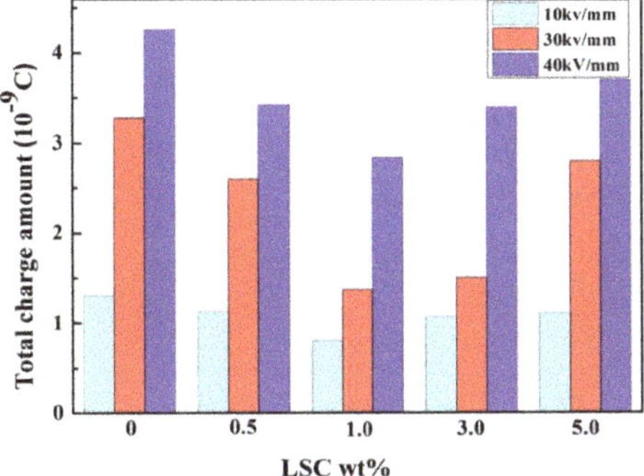

Figure 11. The total amount of charge in the LDPE when the composite with different LSC content acts as a semiconductive layer.

In general, when the composite material with a 1% LSC content is used as a semiconductive layer, the peak value of the depolarization current is the smallest. The depolarization currents have the same tendency at different polarization voltages.

3.6. Space Charge Distribution

The space charge distribution of LDPE under a 10 kV/mm and a 40 kV/mm DC electric field within 30 min at room temperature is shown in Figures 12 and 13. It can be seen from Figure 12a that the accumulation of the homocharge is observed near the cathode and the anode in the LDPE when the semiconductive layer is not added to with LSC. Among them, the heterocharge is derived from the ionization of the crosslinked byproducts and the ionization of the impurities, and the homocharge is derived from the injection of the electrodes. It can be seen from Figures 12c and 13c that there is almost no accumulation of the homo charge at the cathode. However when the content of LSC in the semiconductive layer exceeds 1%, as the LSC content increases, the space charge injection in the LDPE increases; that is, the inhibition effect of the semiconductive layer is weakened, which may be related to the agglomeration of the LSC. It can be inferred that semiconductive materials with an LSC content of 1% can suppress the injection of space charge. Due to the scattering effect at the interface between the nanoparticles and the polymer, the mean free path of electrons is increased and the migration rate of electrons is reduced.

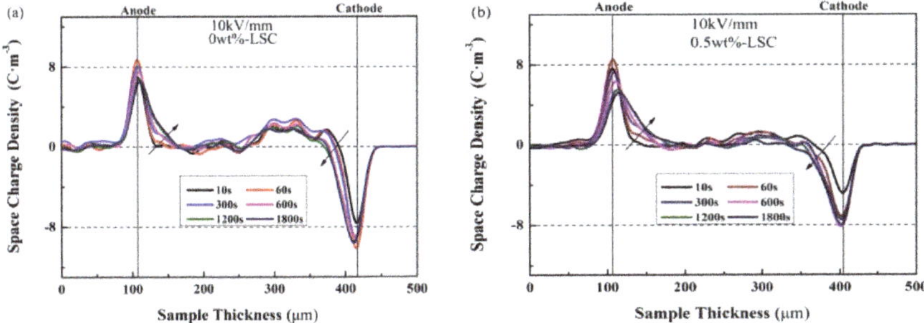

Figure 12. *Cont.*

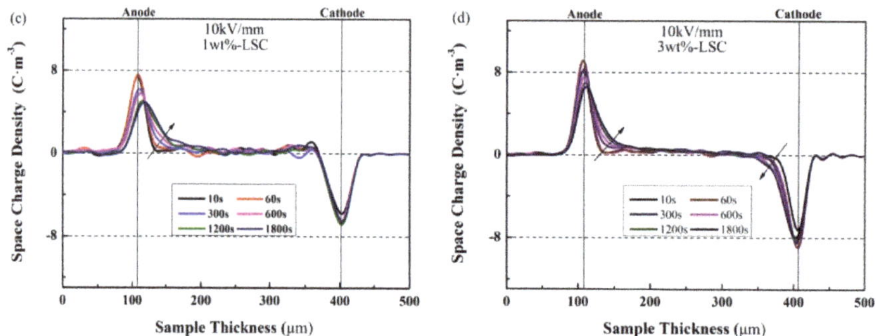

Figure 12. Space charge distribution of LDPE when the composite with different LSC contents acts as a semiconductive layer under 10 kV/mm DC electric field.

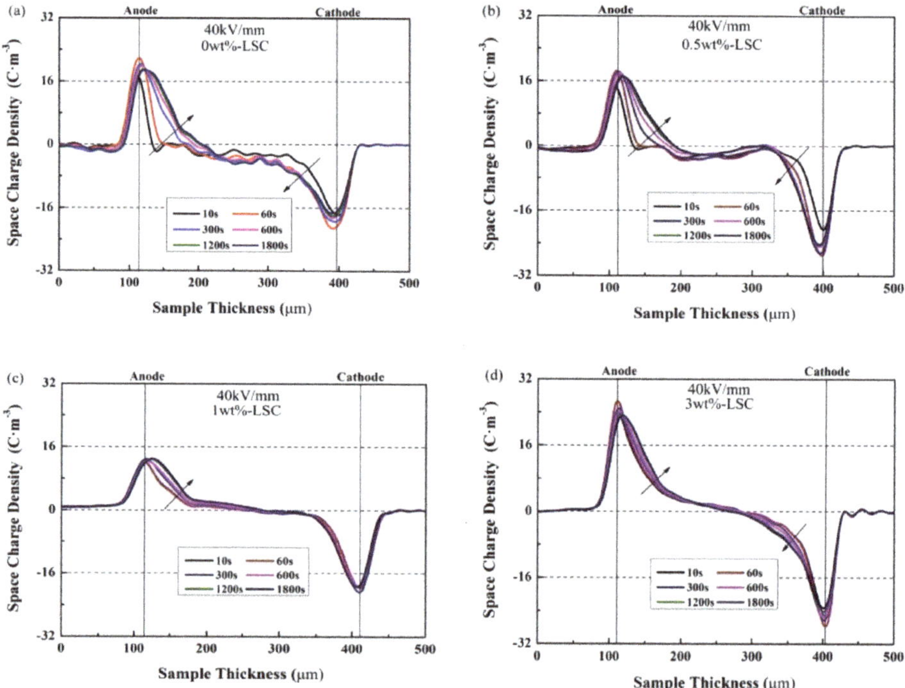

Figure 13. Space charge distribution of LDPE when the composite with different LSC contents acts as a semiconductive layer under 40 kV/mm DC electric field.

When the composite without LSC is used as the semiconductive layer, the maximum charge density near the cathode and anode is 11.46 and 9.37 C/m^3 respectively under a 10 kV/mm DC electric field. After doping by LSC with 1 wt%, the interface charge near the cathode and the anode is reduced to 6.53 and 7.76 C/m^3. The maximum charge density near the two electrodes is 22.15 and 21.36 C/m^3 under a 40 kV/mm DC electric field. When the semiconductive layer is doped with 1 wt% LSC, the interface charge reduced to 12.77 and 20.89 C/m^3, respectively. When the charge is injected from the metal electrode to the insulating layer, it passes through the semiconductive layer, and the charge receives the Coulomb effect of the LSC particles in the semiconducting shielding layer, so that part of the charge cannot be injected into the insulating layer through the semiconductive layer, thus reducing the charge injection in the insulating layer.

4. Discussion

$LaCoO_3$ has a typical perovskite structure. When Sr^{2+} is added into the perovskite lattice to replace La^{3+}, the net electric imbalance will be caused. In order to compensate the net electric imbalance, oxygen vacancy will be generated in the lattice to bring many holes to achieve the charge balance, and oxygen vacancy is allowed to transfer through the perovskite lattice [30]. When the charge is injected from the metal electrode into the insulating layer, it needs to pass through the semiconductive layer. The oxygen vacancy of LSC crystal in the semiconductive layer has electrostatic attraction to the charge, which hinders the movement of the charge, making it difficult for the charge to be injected into the insulating layer through the semiconductive layer, thus reducing the charge injection in the insulating layer.

On the other hand, in ionic crystals, alternating charged plane stacking can generate divergent electrostatic energy, which makes the oxide surface polar. This polar surface is electrostatically unstable, and surface charge must be compensated by surface reconstruction or charged defect accumulation [31]. When the charge is injected from the metal electrode to the insulating layer, it passes through the semiconductive layer. Under the action of electric field, due to the polarity of LSC particle surface, the ions of LSC crystal will move relatively, which will cause polarization, and then lead to the interaction between the polarization field and the charge, thus reducing the charge injection in the insulator. In 1993, Landau proposed that electrons could trap themselves in the deformed lattice [32]. In 2002, Iwanaga et al. observed trapped electrons and holes in PbBr2 crystal [33]. When electrons change from free-form to self-trapped, their mobility will change obviously. As shown in Figure 14, if electrons are injected into the lattice, due to the effect of electrons on the crystal lattice, the surrounding crystal lattice is distorted, causing the positive ions around it to move closer to the electrons, and the negative ions to move far away, which is called a "polarized cloud". The polaron is a combination of electrons and a polarized cloud around it. As the electrons move to drag the surrounding polarized clouds, the mass increases and the migration rate decreases. Lattice deformation can bind electrons, thereby, the injection of electrons from the metal electrode to the insulating layer was suppressed.

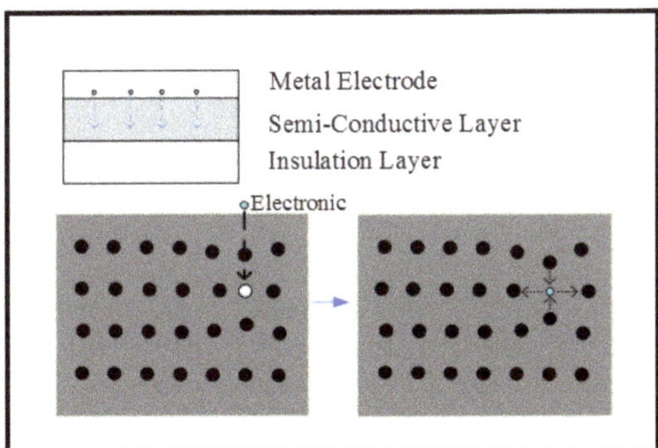

Figure 14. Schematic diagram of electron trapping.

5. Conclusions

In this paper, semiconductive layers with different contents of LSC were prepared by melt blending. The appearance and resistivity of the nanocomposites and their effects on space charge injection of insulating layers were studied. The conclusions are drawn as follows:

1.   When the LSC content in the semiconductive composites is low, the nanoparticles are uniformly dispersed in the matrix, and when the content of the nanoparticles increases, agglomeration occurs.

2. The addition of LSC can suppress the PTC effect of the semiconducting layer. When the LSC content is 1 wt%, the PTC strength of semiconducting shielding layer decreased from 1.47 to 0.99, decreasing by 38.1%. This is because the LSC doped in semiconductive materials is an ionic conductor, and the mobility of carriers increases with the increase in temperature.

3. The experimental results show that when the doping amount of LSC is 1 wt%, the charge amount in the insulating sample is the smallest, which is 0.75×10^{-9}, 1.34×10^{-9}, and 2.75×10^{-9}, respectively, decreasing by 44.4%, 58.9%, and 35.7%. This is because the charge is subjected to the Coulomb force of the LSC particles in the semiconductive layer, which reduces the charge injection from the metal electrode to the insulating layer.

Author Contributions: Conceptualization, C.H. and Y.C.; methodology, H.Y.; software, X.G. and S.Z.; validation, Y.W., G.L. and M.X.; writing—original draft preparation, Y.C.; writing—review and editing, C.H. and Y.C.; funding acquisition, Q.L. and Z.X. All authors read and agreed to the published version of the manuscript.

Funding: This research was funded by [the State Key Laboratory of Advanced Power Transmission Technology] grant number [SGGR0000DWJS1800561] and [State Grid Shandong Electric Power Research Institute] grant number [SGSDDK00KJJS1900329].

Acknowledgments: This research was funded by [the State Key Laboratory of Advanced Power Transmission Technology] grant number [SGGR0000DWJS1800561] and [State Grid Shandong Electric Power Research Institute] grant number [SGSDDK00KJJS1900329].

Conflicts of Interest: The authors declare no conflict of interest.

References

1. Alassi, A.; Bañales, S.; Ellabban, O.; Adam, G.; Maciver, C. HVDC Transmission: Technology Review, Market Trends and Future Outlook. *Renew. Sustain. Energy Rev.* **2019**, *112*, 530–554. [CrossRef]

2. Ohki, Y. Development of XLPE-insulated cable for high-voltage dc ubmarinetransmission line. *IEEE Electr. Insul. Mag.* **2013**, *29*, 65–67. [CrossRef]

3. Zhang, Y.; Zhang, C.; Feng, Y.; Zhang, T.; Chen, Q.; Chi, Q.; Liu, L.; Li, G.; Cui, Y.; Wang, X.; et al. Excellent energy storage performance and thermal property of polymer-based composite induced by multifunctional one-dimensional nanofibers oriented in-plane direction. *Nano Energy* **2019**, *56*, 138–150. [CrossRef]

4. Sun, J.; Rensselaer Polytechnic Institute; Li, M.; Zhang, Z.; Xu, T.; He, J.; Wang, H.; Li, G.; State Grid Corporation of China; China Electric Power Research Institute. Renewable energy transmission by HVDC across the continent: System challenges and opportunities. *CSEE J. Power Energy Syst.* **2017**, *3*, 353–364. [CrossRef]

5. Tiku, D. dc Power Transmission: Mercury-Arc to Thyristor HVdc Valves [History]. *IEEE Power Energy Mag.* **2014**, *12*, 76–96. [CrossRef]

6. Pierri, E.; Binder, O.; Hemdan, N.; Kurrat, M. Challenges and opportunities for a European HVDC grid. *Renew. Sustain. Energy Rev.* **2017**, *70*, 427–456. [CrossRef]

7. Gu, X.; He, S.; Xu, Y.; Yan, Y.; Hou, S.; Fu, M. Partial discharge detection on 320 kV VSC-HVDC XLPE cable with artificial defects under DC voltage. *IEEE Trans. Dielectr. Electr. Insul.* **2018**, *25*, 939–946. [CrossRef]

8. Burns, N.; Eichhorn, R.; Reid, C. Stress controlling semiconductive shields in medium voltage power distribution cables. *IEEE Electr. Insul. Mag.* **1992**, *8*, 8–24. [CrossRef]

9. Fang, Y.; Zhao, J.; Zha, J.-W.; Wang, D.; Dang, Z.-M. Improved stability of volume resistivity in carbon black/ethylene-vinyl acetate copolymer composites by employing multi-walled carbon nanotubes as second filler. *Polymer* **2012**, *53*, 4871–4878. [CrossRef]

10. Kono, A.; Shimizu, K.; Nakano, H.; Goto, Y.; Kobayashi, Y.; Ougizawa, T.; Horibe, H. Positive-temperature-coefficient effect of electrical resistivity below melting point of poly(vinylidene fluoride) (PVDF) in Ni particle-dispersed PVDF composites. *Polymer* **2012**, *53*, 1760–1764. [CrossRef]

11. Dang, Z.-M.; Li, W.-K.; Xu, H.-P. Origin of remarkable positive temperature coefficient effect in the modified carbon black and carbon fiber cofillled polymer composites. *J. Appl. Phys.* **2009**, *106*, 24913. [CrossRef]

12. Fang, Z.; Shang, M.; Hou, X.; Zheng, Y.; Du, Z.; Yang, Z.; Chou, K.-C.; Yang, W.; Wang, Z.L.; Yang, Y. Bandgap alignment of α-CsPbI$_3$ perovskites with synergistically enhanced stability and optical performance via B-site minor doping. *Nano Energy* **2019**, *61*, 389–396. [CrossRef]

13. Guo, H.; Chen, H.; Zhang, H.; Huang, X.; Yang, J.; Wang, B.; Li, Y.; Wang, L.; Niu, X.; Wang, Z.M. Low-temperature processed yttrium-doped SrSnO$_3$ perovskite electron transport layer for planar heterojunction perovskite solar cells with high efficiency. *Nano Energy* **2019**, *59*, 1–9. [CrossRef]

14. Jung, J.-I.; Jeong, H.Y.; Lee, J.; Kim, M.G.; Cho, J. A Bifunctional Perovskite Catalyst for Oxygen Reduction and Evolution. *Angew. Chem. Int. Ed.* **2014**, *53*, 4582–4586. [CrossRef]

15. Kubicek, M.; Cai, Z.; Ma, W.; Yildiz, B.; Hutter, H.; Fleig, J. Tensile Lattice Strain Accelerates Oxygen Surface Exchange and Diffusion in La$_{1-x}$SrxCoO$_{3-\delta}$ Thin Films. *ACS Nano* **2013**, *7*, 3276–3286. [CrossRef] [PubMed]

16. Tanaka, Y.; Chen, G.; Zhao, Y.; Davies, A.; Vaughan, A.S.; Takada, T. Effect of additives on morphology and space charge accumulation in low density polyethylene. *IEEE Trans. Dielectr. Electr. Insul.* **2003**, *10*, 148–154. [CrossRef]

17. Zhang, Y.; Christen, T.; Meng, X.; Chen, J.; Rocks, J. Research progress on space charge characteristics in polymeric insulation. *J. Adv. Dielectr.* **2016**, *6*, 1630001. [CrossRef]

18. Suzuoki, Y.; Ieda, M.; Yoshifuji, N.; Matsukawa, Y.; Han, S.; Fujii, A.; Kim, J.-S.; Mizutani, T. Study of space-charge effects on dielectric breakdown of polymers by direct probing. *IEEE Trans. Electr. Insul.* **1992**, *27*, 758–762. [CrossRef]

19. Kaneko, K.; Mizutani, T.; Suzuoki, Y. Computer simulation on formation of space charge packets in XLPE films. *IEEE Trans. Dielectr. Electr. Insul.* **1999**, *6*, 152–158. [CrossRef]

20. Tanaka, T. Charge transfer and tree initiation in polyethylene subjected to AC voltage stress. *IEEE Trans. Electr. Insul.* **1992**, *27*, 424–431. [CrossRef]

21. Hayase, Y.; Aoyama, H.; Matsui, K.; Tanaka, Y.; Takada, T.; Murata, Y. Space Charge Formation in LDPE/MgO Nano-composite Film under Ultra-high DC Electric Stress. *IEEJ Trans. Fundam. Mater.* **2006**, *126*, 1084–1089. [CrossRef]

22. Takada, T.; Hayase, Y.; Tanaka, Y.; Okamoto, T. Space charge trapping in electrical potential well caused by permanent and induced dipoles for LDPE/MgO nanocomposite. *IEEE Trans. Dielectr. Electr. Insul.* **2008**, *15*, 152–160. [CrossRef]

23. Lv, Z.; Wang, X.; Wu, K.; Chen, X.; Cheng, Y.; Dissado, L.A. Dependence of charge accumulation on sample thickness in Nano-SiO$_2$ doped IDPE. *IEEE Trans. Dielectr. Electr. Insul.* **2013**, *20*, 337–345. [CrossRef]

24. Zhang, L.; Zhou, Y.; Huang, M.; Sha, Y.; Tian, J.; Ye, Q. Effect of nanoparticle surface modification on charge transport characteristics in XLPE/SiO$_2$ nanocomposites. *IEEE Trans. Dielectr. Electr. Insul.* **2014**, *21*, 424–433. [CrossRef]

25. Zhou, Y.; Hu, J.; Dang, B.; He, J. Effect of different nanoparticles on tuning electrical properties of polypropylene nanocomposites. *IEEE Trans. Dielectr. Electr. Insul.* **2017**, *24*, 1380–1389. [CrossRef]

26. Wei, Y.; Liu, M.; Wang, J.; Li, G.; Hao, C.; Lei, Q. Effect of Semi-Conductive Layer Modified by Magnetic Particle SrFe$_{12}$O$_{19}$ on Charge Injection Characteristics of HVDC Cable. *Polymers* **2019**, *11*, 1309. [CrossRef]

27. Rousseau, S.; Loridant, S.; Delichère, P.; Boreave, A.; Deloume, J.; Vernoux, P. La$_{(1-x)}$SrxCo$_{1-y}$Fe$_y$O$_3$ perovskites prepared by sol–gel method: Characterization and relationships with catalytic properties for total oxidation of toluene. *Appl. Catal. B* **2009**, *88*, 438–447. [CrossRef]

28. Grinberg, I.; West, D.V.; Torres, M.; Gou, G.; Stein, D.M.; Wu, L.; Chen, G.; Gallo, E.M.; Akbashev, A.R.; Davies, P.K.; et al. Perovskite oxides for visible-light-absorbing ferroelectric and photovoltaic materials. *Nature* **2013**, *503*, 509–512. [CrossRef]

29. Lee, J.J.; Oh, M.Y.; Nahm, K.S. Effect of Ball Milling on Electrocatalytic Activity of Perovskite La0.6Sr0.4CoO3-δApplied for Lithium Air Battery. *J. Electrochem. Soc.* **2015**, *163*, A244–A250. [CrossRef]

30. Glisenti, A.; Vittadini, A. On the Effects of Doping on the Catalytic Performance of (La,Sr)CoO$_3$. A DFT Study of CO Oxidation. *Catalysts* **2019**, *9*, 312. [CrossRef]

31. Setvin, M.; Reticcioli, M.; Poelzleitner, F.; Hulva, J.; Schmid, M.; Boatner, L.A.; Franchini, C.; Diebold, U. Polarity compensation mechanisms on the perovskite surface KTaO$_3$(001). *Science* **2018**, *359*, 572–575. [CrossRef]

32. Landau, L. The motion of electrons in a crystal lattice. *Phys. Zeits. Sowjetunion.* **1993**, *3*, 664–665.
33. Iwanaga, M.; Hayashi, T.; Tanaka, J. Society, Azuma, Masanobu Shirai, and KoichiroSefrep detos and hols in PbBr: Cysal. *Am. Phys.* **2002**, *65*, 214306.

 *applied sciences*

Article

Optimal Scheduling of Large-Scale Wind-Hydro-Thermal Systems with Fixed-Head Short-Term Model

Thang Trung Nguyen [1], Ly Huu Pham [1], Fazel Mohammadi [2,*] and Le Chi Kien [3]

[1]   Power System Optimization Research Group, Faculty of Electrical and Electronics Engineering,
    Ton Duc Thang University, Ho Chi Minh City 700000, Vietnam; nguyentrungthang@tdtu.edu.vn (T.T.N.);
    phamhuuly@tdtu.edu.vn (L.H.P.)
[2]   Electrical and Computer Engineering (ECE) Department, University of Windsor, Windsor,
    ON N9B 1K3, Canada
[3]   Faculty of Electrical and Electronics Engineering, Ho Chi Minh City University of Technology and Education,
    Ho Chi Minh City 700000, Vietnam; kienlc@hcmute.edu.vn
*   Correspondence: fazel@uwindsor.ca or fazel.mohammadi@ieee.org

Received: 27 March 2020; Accepted: 21 April 2020; Published: 24 April 2020

Abstract: In this paper, a Modified Adaptive Selection Cuckoo Search Algorithm (MASCSA) is proposed for solving the Optimal Scheduling of Wind-Hydro-Thermal (OSWHT) systems problem. The main objective of the problem is to minimize the total fuel cost for generating the electricity of thermal power plants, where energy from hydropower plants and wind turbines is exploited absolutely. The fixed-head short-term model is taken into account, by supposing that the water head is constant during the operation time, while reservoir volume and water balance are constrained over the scheduled time period. The proposed MASCSA is compared to other implemented cuckoo search algorithms, such as the conventional Cuckoo Search Algorithm (CSA) and Snap-Drift Cuckoo Search Algorithm (SDCSA). Two large systems are used as study cases to test the real improvement of the proposed MASCSA over CSA and SDCSA. Among the two test systems, the wind-hydro-thermal system is a more complicated one, with two wind farms and four thermal power plants considering valve effects, and four hydropower plants scheduled in twenty-four one-hour intervals. The proposed MASCSA is more effective than CSA and SDCSA, since it can reach a higher success rate, better optimal solutions, and a faster convergence. The obtained results show that the proposed MASCSA is a very effective method for the hydrothermal system and wind-hydro-thermal systems.

Keywords: Cuckoo Search Algorithm (CSA); Fixed-Head Short-Term Model; Hydrothermal System; Optimal Scheduling of Wind-Hydro-Thermal System (OSWHTS)

---

1. Introduction

Short-term hydrothermal scheduling considers optimization horizon from one day to one week, involving the hour-by-hour generation planning of all generating units in the hydrothermal system, so that the total generation fuel cost of thermal units is minimized, while satisfying all constraints from hydropower plants, including hydroelectric power plant constraints, such as water discharge limits, volume reservoir limits, continuity water, generation limits, and thermal power plant constraints, including prohibited operating zone and generation limits. There is a fact that the load demand changes cyclically over one day or one week, and varies corresponding to the short-term scheduling horizon, which is in a range from one day to one week. A set of beginning conditions, consisting of initial and final reservoir volumes for the scheduling horizon, inflow into the reservoir, and the water amount to be used for the scheduling horizon, is assumed to be known. During the scheduling generation process, it is necessary to consider the capacity of the reservoir and inflow once they have significant impacts on the water head variations, and lead to being represented by different hydro models. In this paper, a fixed-head short-term hydrothermal scheduling with reservoir volume constraints is considered. The reservoir water head is supposed to be fixed during the scheduling horizon [1]. Therefore, the water discharge is still the second-order function of hydro generation and given coefficients. The total amount of water is not required to be calculated and constrained. However, the initial and final values of the Reservoir Volume Should Be met with the optimal operation of the hydrothermal system. The capacity of the reservoir to contain water during the operation must be observed and followed by the constrained values, such as minimum volume corresponding to the deadhead and maximum volume corresponding to the highest head. Moreover, the continuity of water is always constrained at each subinterval over the scheduling horizon. Other issues related to power transmission lines, such as power balance and power losses, are also taken into account for most test systems.

The problem has been studied so far and obtained many intentions from researchers. Several algorithms, such as Gradient Search Algorithm (GSA) [2], Newton–Raphson Method (NRM) [3], Hopfield Neural Networks (HNN) [4], Simulated Annealing Algorithm (SAA) [5], Evolutionary Programming Algorithm (EPA) [6–8], Genetic Algorithm (GA) [9], modified EPA (MEPA) [10], Fast Evolutionary Programming Algorithm (FEPA) [10], Improved FEPA (IFEPA) [10], Hybrid EPA (HEPA) [11], Particle Swarm Optimization (PSO) [12], Improved Bacterial Foraging Algorithm (IBFA) [13], Self-Organization Particle Swarm Optimization (SOPSO) [14], Running IFEPA (RIFEPA) [15], Improved Particle Swarm Optimization (IPSO) [16,17], Clonal Selection Optimization Algorithm (CSOA) [18], Full Information Particle Swarm Optimization (FIPSO) [19], One-Rank Cuckoo Search Algorithm with the applications of Cauchy (ORCSA-Cauchy) and Lévy distribution (ORCSA-Lévy) [20], Cuckoo Search Algorithm with the applications of Gaussian distribution (CSA-Gauss), Cauchy distribution (CSA-Cauchy), and Lévy distribution (CSA-Lévy) [21], Adaptive Cuckoo Search Algorithm (ACSA) [22], Improved Cuckoo Search Algorithm (ICSA) [23], Modified Cuckoo Search Algorithm (MCSA) [24], and Adaptive Selective Cuckoo Search Algorithm (ASCSA) [24] have been applied to solve the problem of hydrothermal scheduling. Almost all of the above-mentioned methods are mainly meta-heuristic algorithms, excluding GSA and NRM. Regarding the development history, GSA and NMR are the oldest methods, with the worst capabilities to deal with constraints and finding high-quality parameters of the problem, and they are applied for hydropower generation function with the piecewise linear form or polynomial approximation form. GSA cannot deal with the systems with complex constraints and also the systems with a large number of constraints and variables. NRM seems to be more effective than GSA when applied to systems where the approximation of the hydro generation cannot be performed. However, this method is fully dependent on the scale of the Jacobian matrix and the capability of taking the partial derivative of the Jacobian matrix with respect to each variable. On the contrary to GSA and NRM, population-based metaheuristic algorithms are successfully applied for solving the complicated problem. Among those methods, SAA and GA are the oldest methods and found low-quality solutions for hydropower plants and thermal power plants. Differently, PSO and EPA variants are more effective in reaching better solutions with faster speed. The improved versions of EPA are not verified, while they were claimed to be much better than conventional EPA. Only one-thermal and one-hydropower plant system and quadratic fuel cost function is employed as the case study for running those methods. In order to improve the conventional PSO successfully, weight factor [16] and constriction factor [17] are respectively used to update new velocity and new position. The improvement also leads to an optimal solution with shorter execution time, but the two research studies report an invalid optimal solution, since the water discharge violates the lower limit. In [19], the new version of the updated velocity of the FIPSO is proposed and tested on a system. However, the method reports an invalid solution violating the lower limit. IBFA [13] also shows an invalid optimal solution with more water than availability. CSOA is demonstrated to be stronger than GA, EP, and Differential Evolution (DE) for this problem. CSA variants [20–24] are developed for the problem and reached better results. Different distributions are tested to find the most appropriate one as compared to original distribution, which is Lévy distribution. Cauchy and Gaussian distributions also result in the same best solution for the system with four hydropower plants and one thermal power plant, but the two distributions cope with a low possibility of finding the best solution.

In recent years, wind energy has been considered as a power source, together with conventional power plants, to supply electricity to loads. The optimal scheduling of thermal power plants and wind turbines is successfully solved using the Artificial Bee Colony Algorithm (ABCA) [25] and Wait-And-See Algorithm (WASA) [26]. Then, the wind-thermal system is expanded by integrating one more conventional power source, which is a hydropower plant, leading to the wind-hydro-thermal system. The optimal scheduling of the wind-hydro-thermal system is performed using different metaheuristic algorithms, such as Nondominated Sorting Genetic Algorithm-III (NSGA-III) [27], Multi-Objective Bee Colony Optimization Algorithm (MOBCOA) [28], Distributionally Robust Hydro-Thermal-Wind Economic Dispatch (DR-HTW-ED) method [29], nonlinear and dynamic Optimal Power Flow (OPF) method [30], Modified Particle Swarm Optimization (MPSO) [31], Mixed Binary and Real Number Differential Evolution (MBRNDE) [32], Mixed-Integer Programming (MIP) [33], Two-Stage Stochastic Programming Model Method (TSSPM) [34], and Sine Cosine Algorithm (SCA) [35]. In general, almost all applied methods are meta-heuristic algorithms and the purpose of those studies is to demonstrate the highly successful constraint handling capability of the applied metaheuristic algorithms, rather than showing high-quality solution searching capability.

In this paper, wind farms, together with the hydrothermal system, are considered to supply electricity to loads, in which the fixed-head short-term hydrothermal system is investigated. The objective of the Optimal Scheduling of Wind-Hydro-Thermal System (OSWHTS) problem is to minimize total electricity generation fuel cost of thermal power plants in a day, subject to the wind farms, reservoirs, and thermal units' constraints. In the fixed-head short-term model, water discharge is a second-order equation, with respect to the power output of the hydropower plant. In addition, hydraulic constraints are discharge limits, reservoir volume limits, initial reservoir volume, and end reservoir volume. In order to solve the OSWHTS problem successfully and effectively, a Modified Adaptive Selection Cuckoo Search Algorithm (MASCSA) is proposed by applying two new modifications on the Adaptive Selection Cuckoo Search Algorithm (ASCSA), which was first developed in [24]. In addition, other metaheuristic algorithms are implemented for comparisons. The implemented algorithms are CSA [36] and SDCSA [37]. CSA was first introduced by Yang and Deb in 2009 [36], and it has been widely applied for different optimization problems in electrical engineering. However, CSA is indicated to be less effective for large and complicated problems [24,37]. Hence, SDCSA and ASCSA are proposed. SDCSA is applied only for benchmark functions, while ASCSA is more widely applied for three complicated hydrothermal scheduling problems. ASCSA is superior to many existing meta-heuristic algorithms, such as GA, DE, and other CSA variants. ASCSA is an improved version of CSA, by implementing two more modifications, including a new selection technique and an adaptive mutation mechanism. ASCSA can reach high performance, but it suffers from long simulation time, due to the selection of mutation factor and threshold. Thus, in this paper, two new modifications, including setting the mutation factor to one and proposing a new condition for replacing the threshold, are applied.

The novelties of the paper are the integration of wind turbines and the fixed-head short-term hydrothermal system and a proposed CSA, called MASCSA. Thanks to the novelties, the main contributions of the study are the most appropriate selection of control variables for the optimal scheduling of the wind-hydro-thermal system, the effective constraint handling method, and the high performance proposed MASCSA method.

The rest of the paper is organized as follows. The formulation of the OSWHTS problem is given in Section 2. The details of the proposed method are described in Section 3. The search process of

MASCSA for the OSWHTS problem is presented in Section 4. The comparison results of the two test systems are given in Section 5. Finally, the conclusions are summarized in Section 6.

2. Formulation of Optimal Scheduling of Wind-Hydro-Thermal System

In this section, the optimal scheduling problem of the wind-hydro-thermal system with the fixed-head short-term model of a hydropower plant is mathematically expressed considering the objective function and constraints. A typical wind-hydro-thermal system is shown in Figure 1. From the figure, N_h hydropower plants, N_t thermal power plants, and N_w turbines in a wind farm are generating and supplying electricity to loads via different buses. The purpose of the system is to minimize the total electricity generation cost of N_t thermal power plants, considering the available water in reservoirs and the intermittent nature of wind power. The cost of generated power by hydropower plants and the wind farm is neglected, but all constraints from the plants are supervised. The objective function and all constraints can be mathematically formulated as follows:

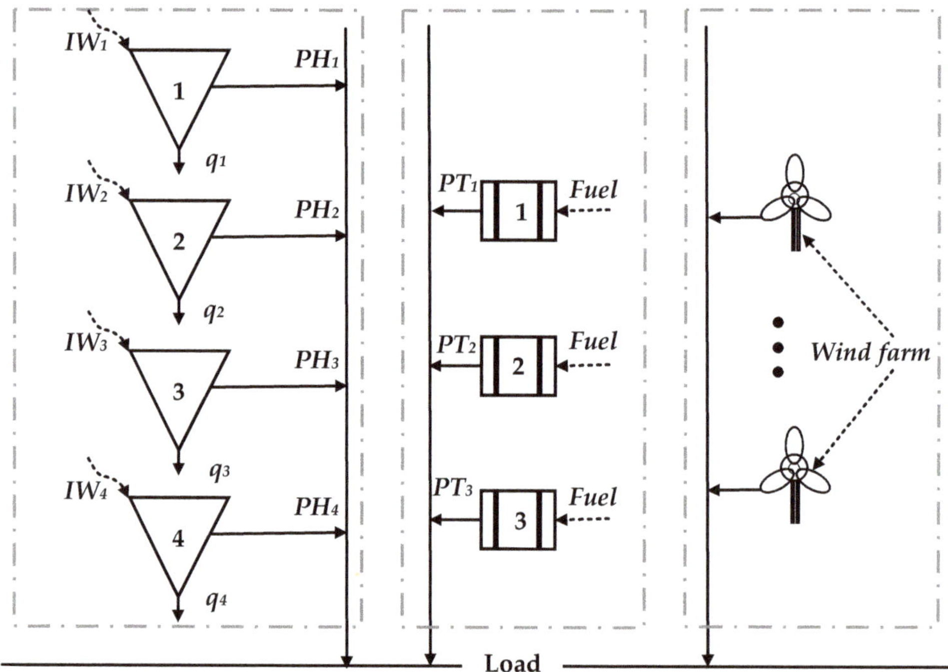

Figure 1. A typical wind-hydro-thermal system.

2.1. Total Electricity Generation Fuel Cost Reduction Objective

Total fuel cost for generating electricity from all thermal power plants is considered as a major part that needs to be minimized as much as possible. The objective is shown as follows:

$$TFC = \sum_{j=1}^{N_s} \sum_{tp=1}^{N_t} t_i \left(k_{tp} + m_{tp} PT_{tp,i} + n_{tp,i} \left(PT_{tp,i} \right)^2 + \left| \alpha_{tp} \times \sin\left(\beta_{tp} \times \left(PT_{tp,min} - PT_{tp,i} \right) \right) \right| \right) \tag{1}$$

2.2. Set of Constraints and Wind Model

2.2.1. Constraints from Hydropower Plants

Hydropower plants are constrained by limits of reservoirs, turbines, and generators. The detail is expressed as follows:

Water Balance Constraint: The reservoir volume at the ith considered subinterval is always related to the volume of previous subinterval, water inflow, and water discharge. All the parameters must be supervised so that the following equality is exactly met.

$$RV_{hp,i-1} - RV_{hp,i} + WI_{hp,i} - Q_{hp,i} = 0, \; i = 1, 2, \ldots, N_s \tag{2}$$

Note that $RV_{hp,i-1}$ is equal to $V_{hp,0}$, if $i = 1$, and $RV_{hp,i}$ is equal to RV_{hp,N_s}, if $i = N_s$.

Initial and Final Volumes Constraints: $V_{hp,0}$ and V_{hp,N_s} in constraint (2) should be equal to two given parameters, as shown in the model below.

$$RV_{hp,0} = RV_{hp,start} \tag{3}$$

$$RV_{hp,N_s} = V_{hp,end} \tag{4}$$

For each operating day, initial volume, $RV_{hp,start}$, and final volume, $RV_{hp,end}$, of each reservoir are required to be always exactly met.

Reservoir Operation Limits: Water volume of reservoirs must be within the upper and lower limits in order to assure that the water head is always in operation limits. Therefore, the following inequality is an important constraint.

$$RV_{hp,min} \leq RV_{hp,i} \leq RV_{hp,max}, \; \begin{cases} hp = 1, 2, \ldots, N_h \\ i = 1, 2, \ldots, N_s \end{cases} \tag{5}$$

Limits of Discharge Through Turbines: Turbines of each hydropower plant is safe, if the water discharge through them does not exceed the limits. Both upper and lower limits have a huge meaning for the safety and stable operation of turbines. Thus, the following constraints are considered.

$$q_{hp,min} \leq q_{hp,i} \leq q_{hp,max}, \; \begin{cases} hp = 1, 2, \ldots, N_h \\ i = 1, 2, \ldots, N_s \end{cases} \tag{6}$$

where $q_{hp,i}$ is determined as follows:

$$q_{hp,i} = x_{hp} + y_{hp}PH_{hp,i} + z_{hp}\left(PH_{hp,i}\right)^2 \tag{7}$$

In addition, the total discharge of each subinterval is determined as follows:

$$Q_{hp,i} = t_i q_{hp,i} \tag{8}$$

Limits of Hydropower Plant Generators: The power generation of each hydropower plant must follow the inequality below, to assure the safe operation of generators all the time.

$$PH_{hp,min} \leq PH_{hp,i} \leq PH_{hp,max}, \; \begin{cases} hp = 1, 2, \ldots, N_h \\ i = 1, 2, \ldots, N_s \end{cases} \tag{9}$$

2.2.2. Constraint of Thermal Power Plant

It is supposed that thermal power plants have plentiful fossil fuel and their energy is not constrained. However, thermal power plant generators have to satisfy physical limits similar to generators of hydropower plants. Namely, the power generation is limited as follows:

$$PT_{tp,min} \leq PT_{tp,i} \leq PT_{tp,max}, \quad \begin{cases} tp = 1, \, 2, \ldots, \, N_t \\ i = 1, \, 2, \ldots, N_s \end{cases} \tag{10}$$

2.2.3. Constraints of Power Systems

Power systems require the balance between the generated and consumed power for the stable voltage and frequency in power systems [38–43]. The power generation of all hydropower plants and thermal power plants, and power consumed by load and lines must follow the equality below:

$$\sum_{tp=1}^{N_t} PT_{tp,i} \sum_{hp=1}^{N_h} PH_{hp,i} + \sum_{w=1}^{N_w} PW_{w,i} - P_{L,i} - P_{TL,i} = 0 \tag{11}$$

2.2.4. Modeling of Wind Uncertainty

Basically, electricity power from wind turbines is highly dependent on wind speed. The operation characteristics of a typical wind turbine are shown in Figure 2. For the figure, wind turbines cannot generate electricity when the wind speed is lower than WV_{in} and higher than WV_{out}. The generated power by wind turbines, shown in Figure 2, can be also formulated as follows [43,44]:

$$PW_w = \begin{cases} 0, & (WV_w < WV_{in} \text{ and } WV_w > WV_{out}) \\ \frac{(WV_w - WV_{in})}{(WV_r - WV_{in})} \times PW_{w,rate}, & (WV_{in} \leq WV_w \leq WV_r) \\ PW_{w,r}, & (WV_r \leq WV_w \leq WV_{out}) \end{cases} \tag{12}$$

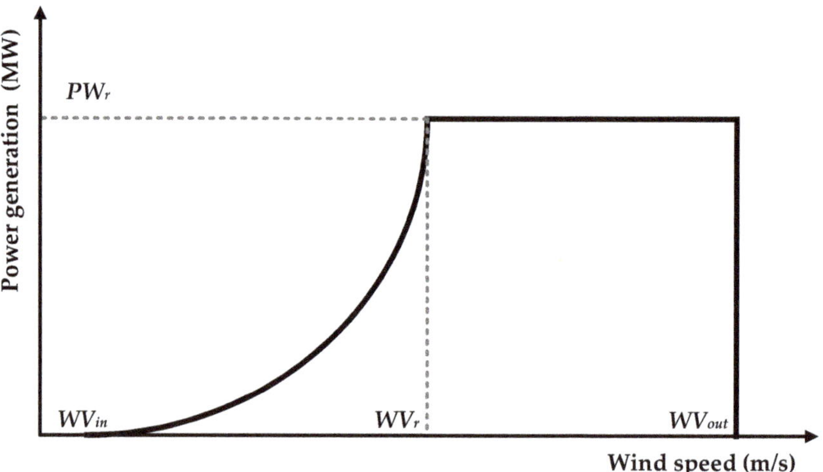

Figure 2. A typical wind turbine characteristic.

3. The Proposed Method

3.1. Conventional Cuckoo Search Algorithm (CSA)

CSA is comprised of two techniques for updating new solutions. The first technique is based on Lévy flights to expand searching space considering very large step sizes. On the contrary, the second technique narrows searching space nearby current solutions, using a mutation operation similar to that in the DE algorithm. Due to different strategies, the first technique is called the exploration phase, whereas the second technique is known as the exploitation phase. The exploration phase is mathematically expressed as follows:

$$So_s^{new} = So_s + \alpha \times (So_s - So_{Gbest}) \otimes \text{Lévy}(\beta) \tag{13}$$

where α is the positive scale factor, which can be selected within the range of 0 and 1; Lévy(β) is the Lévy distribution function [21], and So_{Gbest} is the best solution of the previous iteration.

The exploitation phase can be mathematically expressed as the following mutation technique:

$$So_s^{new} = \begin{cases} So_s + \delta \times (So_1 - So_2), & rd_s < MF \\ So_s, & otherwise \end{cases} \tag{14}$$

where So_1 and So_2 are two randomly generated solutions from the current solutions, rd_s is a randomly generated number within zero and 1, and MF is the mutation factor, which is selected within the range of 0 and 1.

In the exploitation phase, there is a possibility that new solutions cannot be updated, i.e., new solutions and old solutions can be the same. This is particularly the case, given that the mutation factor, MF, is selected to be close to zero, and therefore the possibility that the phenomenon happens is very high. Additionally, it is obvious that new solutions are absolutely updated, if MF is selected to be close to 1.0. Consequently, the searching performance of CSA is highly dependent on the most appropriate value of MF.

3.2. Modified Adaptive Selective Cuckoo Search Algorithm (MASCSA)

The main shortcomings of CSA are indicated in [24], by presenting and analyzing the selection mechanism and mutation mechanism. The two main shortcomings are to miss promising solutions due to the selection mechanism and generate new solutions with low quality, due to the same updated step size of the mutation mechanism. As a result, two modifications are proposed to be the new selection mechanism and the adaptive mutation mechanism. The selection mechanism and the adaptive mutation mechanism are presented in detail as follows:

3.2.1. New Selection Mechanism (NSM)

The selection mechanism in [24] is proposed to retain better solutions in the old and new solution sets. Thus, before implementing the selection between new and old solutions, the old and new solution sets with twice the population are grouped into one. Then, the fitness function is used to sort solutions from the best one to the worst one. Finally, the first population is retained and another one is abandoned.

3.2.2. Adaptive Mutation Mechanism (AMM)

AMM in [24] is applied to use two different sizes of the updated step. In Equation (14), only the step with the deviation between two random solutions is applied. Consequently, the mechanism applies two different sizes for each considered solution, in which the small step size is established by using two solutions, and the large step size is calculated by using four different solutions. The small size and the large size support the formation of new solutions, as shown in the following equations:

$$So_s^{new} = So_s + \delta \times (So_1 - So_2) \tag{15}$$

$$So_s^{new} = So_s + \delta \times (So_1 - So_2) + \delta \times (So_3 - So_4) \tag{16}$$

However, ASCSA has still applied the condition of the comparison between rd_s and MF, shown in Equation (14). Thus, either Equation (15) or Equation (16) is not used if rd_s is higher than MF. Clearly, there is a high possibility that new solutions are not generated if MF is set to close to zero. In order to avoid this shortcoming, MF is set to one in the proposed MASCSA method.

Furthermore, in order to determine the use of either Equation (15) or Equation (16), ASCSA has applied a condition much dependent on a high number of selections. A ratio of fitness function of each considered solution to the fitness function of the best solution is calculated and then the ratio is compared to a threshold, which is suggested to be 10^{-5}, 10^{-4}, 10^{-3}, 10^{-2}, and 10^{-1}. If the ratio is less than the threshold, Equation (15) is used. Otherwise, Equation (16) is selected. Clearly, the condition is time-consuming, due to the selection of five values for the threshold. Consequently, in order to tackle the main disadvantage of ASCSA, a modified adaptive mutation mechanism is proposed and shown in the next section.

3.2.3. The Modified Adaptive Mutation Mechanism (MAMM)

In the MAMM, the adaptive mutation mechanism in [24] is applied, together with a proposed condition for determining the use of small size or large size in Equations (15) and (16). The fitness function of each solution is determined and defined as FF_s. The fitness function is used to calculate the effective index of each solution and the average effective index of the solutions. The effective index of the sth solution, EI_s, and the average effective index of the whole population, EI_a, are calculated as follows:

$$EI_s = FF_{best}/FF_s \tag{17}$$

$$EI_a = FF_{best}/FF_a \tag{18}$$

where FF_{best} and FF_a are the fitness function of the best solution and the average fitness function of the whole population. In the case that the effective index of the sth solution is less than that of the whole population, the sth solution is still far from the so-far best solution and small size should be used for the sth solution. On the contrary, the sth solution may be close to the so-far best solution and the large size is preferred. In summary, the modified adaptive mutation mechanism can be implemented by the five following steps:

Step 1: Set mutation factor MF to one
Step 2: Calculate the fitness function of the sth solution, FF_s and determine the lowest one, FF_{best}
Step 3: Calculate the mean fitness function of all current solutions, FF_a
Step 4: Calculate EI_s and EI_a using Equations (17) and (18)
Step 5: Compare EI_s and EI_a

If $EI_s < EI_a$, apply Equation (15) for the sth solution.

Otherwise, apply Equation (16) for the sth solution.

Using the AMM [34], ASCSA can jump to promising search zones with appropriate step size, as shown in Equations (15) and (16). However, the condition for applying either Equation (15) or Equation (16) is time-consuming, due to the many values of threshold, including 10^{-5}, 10^{-4}, 10^{-3}, 10^{-2}, and 10^{-1}. In addition, the mutation factor is also set to the range from 0.1 to 1.0 with ten values. Therefore, it should try (5×10) = 50 values for the ASCSA. This becomes a serious issue of ASCSA in finding the best solution. Therefore, the application of the new condition can enable MASCSA to reach high performance, but the shortcomings of the time-consuming manner can be solved easily.

4. The Application of the Proposed MASCSA Method for OSWHT Problem

4.1. Decision Variables Selection

Solution methods can be applied for an optimization problem with the first step of determining decision variables, which are included in each candidate solution. In the problem, the decision variables are selected to be as follows:

1. Reservoir volume of all hydropower plants at the first subinterval to the $(N_s - 1)^{th}$ subinterval: $V_{hp,i}$, where $hp = 1, \ldots, N_h$ and $i = 1, \ldots, N_s$.
2. Power generation of the first $(N_t - 1)$ thermal power plants for all subinterval: $PT_{tp,i}$, where $tp = 1, \ldots, N_t - 1$ and $i = 1, \ldots, N_s$.

4.2. Handling Constraints of Hydropower Plants

From the constraint of water balance in Equation (2), the total discharge of each subinterval is obtained as follows:

$$Q_{hp,i} = V_{hp,i-1} - V_{hp,i} + WI_{hp,i}, \ i = 1, 2, \ldots, N_s \tag{19}$$

Then, the discharge of each hour is determined using Equation (8), as follows:

$$q_{hp,i} = \frac{Q_{hp,i}}{t_i}, \ \begin{cases} hp = 1, 2, \ldots, N_h \\ i = 1, 2, \ldots, N_s \end{cases} \tag{20}$$

As a result, the power generation of hydropower plants can be found using Equation (7).

4.3. Handling Power Balance Constraint

From the power balance constraint shown in (11), the power generation of the N_t^{th} thermal power plant is determined as follows:

$$PT_{N_t,i} = P_{L,i} + P_{TL,i} - \sum_{tp=1}^{N_t-1} PT_{tp,i} - \sum_{hp=1}^{N_h} PH_{hp,i} - \sum_{w=1}^{N_w} PW_{w,i} \tag{21}$$

4.4. Fitness Function

The fitness function of each solution is determined to evaluate the quality of the solution. Therefore, the total electricity fuel cost of all thermal power plants and all constraints that have the possibility to be violated are the major terms of the fitness function. As shown in Section 4.1, reservoir volume and power generation of the first $(N_t - 1)$ thermal power plants are the decision variables. Hence, they never violate the limits. However, the discharge of each hour and power generation of hydropower plants, and the last thermal power plant, have a high possibility of violating both the upper and lower limits. Derived from the meaning, the solution quality evaluation function is established as follows:

$$FF_s = TFC + PF_1 \times \sum_{hp=1}^{N_h} \sum_{i=1}^{N_s} \Delta q_{hp,i}^2 + PF_2 \times \sum_{hp=1}^{N_h} \sum_{i=1}^{N_s} \Delta PH_{hp,i}^2 + PF_3 \times \sum_{i=1}^{N_s} \Delta PT_{N_t,i}^2 \tag{22}$$

where PF_1, PF_2, and PF_3 are the penalty factors corresponding to the violation of discharge, power generation of hydropower, and power generation of the last thermal power plant, respectively. $\Delta q_{hp,i}$, $\Delta PH_{hp,i}$, and $\Delta PT_{N_t,i}$ are the penalty terms of discharge, power generation of hydropower plants, and power generation of the last thermal power plants. The penalty terms in Equation (22) are determined as follows:

$$\Delta q_{hp,i} = \begin{cases} \left(q_{hp,i} - q_{hp,max}\right), & q_{hp,i} > q_{hp,max} \\ \left(q_{hp,min} - q_{hp,i}\right), & q_{hp,i} < q_{hp,min} \\ 0, & otherwise \end{cases} \tag{23}$$

$$\Delta PH_{hp,i} = \begin{cases} \left(PH_{h_p,i} - PH_{h_p,max}\right), & PH_{h_p,i} > PH_{h_p,max} \\ \left(PH_{h_p,min} - PH_{hp,i}\right), & PH_{h_p,i} < PH_{h_p,min} \\ 0, & otherwise \end{cases} \tag{24}$$

$$\Delta PT_{Nt,i} = \begin{cases} \left(PT_{Nt,i} - PT_{Nt,max}\right), & PT_{Nt,i} > PT_{Nt,max} \\ \left(PH_{Nt,min} - PT_{Nt,i}\right), & PT_{Nt,i} < PT_{Nt,min} \\ 0, & otherwise \end{cases} \tag{25}$$

4.5. The Whole Application Procedure of MASCSA for OSWHT Problem

The whole solution process of the optimal scheduling of the wind-hydro-thermal system with the fixed-head short-term model is described in Figure 3, as follows:

Step 1: Set values to P_s and $Iter^{max}$

Step 2: Randomly initialize So_s (s=1, ... , P_s) within the lower and upper bounds

Step 3: Calculate $PW_{w,i}$ using Equation (12)

Step 4: Calculate $Q_{hp,i}$, $q_{hp,i}$ and $PH_{hp,i}$ using Equations (19), (20), and (7).

Step 5: Calculate $PT_{Nt,i}$ using Equation (21)

Step 6: Calculate the fitness function using Equations (22)–(25)

Step 7: Determine So_{Gbest} and set current iteration to 1 ($Iter$=1)

Step 8: Generate new solutions using Equation (13) and correct the solutions

Step 9: Calculate $Q_{hp,i}$, $q_{hp,i}$, and $PH_{hp,i}$ using Equation (19), (20), and (7).

Step 10: Calculate $PT_{Nt,i}$ using Equation (21)

Step 11: Calculate fitness function using Equations (22)–(25)

Step 12: Compare FF_s^{new} and FF_s to keep better solutions

Step 13: Generate new solutions using MAMM and correct the solutions

Step 14: Calculate $Q_{hp,i}$, $q_{hp,i}$, and $PH_{hp,i}$ using Equations (19), (20), and (7).

Step 15: Calculate $PT_{Nt,i}$ using Equation (21)

Step 16: Calculate fitness function using Equations (22)–(25)

Step 17: Apply NSM in Section 3.2.1.

Step 18: Determine So_{Gbest}

Step 19: If $Iter$= $Iter^{max}$, stop the solution searching algorithm. Otherwise, set $Iter$= $Iter$+1 and go back to Step 8

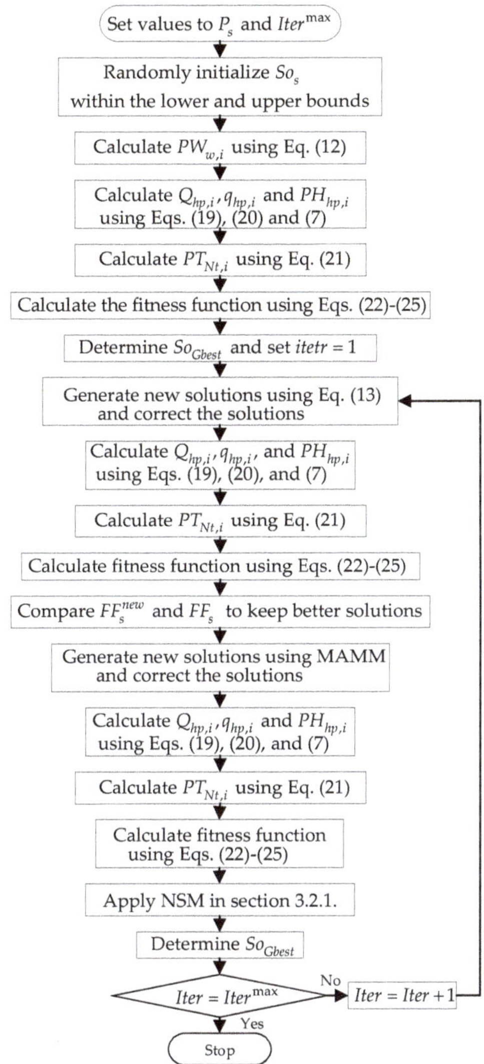

Figure 3. The flowchart for implementing MASCSA for OSWHT problem.

5. Numerical Results

In this section, the performance of the proposed MASCSA is investigated by comparing the results of the proposed method to those from other implemented methods, such as CSA and SDCSA. Two test systems are employed as follows:

1. Test System 1: Four hydropower plants and four thermal power plants with valve effects are optimally scheduled over one day with twenty-four one-hour subintervals. The data of the system are modified from Test System 1 in [7] and also reported in Tables A1–A3 in the Appendix A.
2. Test System 2: Four hydropower plants, four thermal power plants, and two wind farms with the rated power of 120 MW and 80 MW are optimally scheduled over one day with twenty-four

one-hour subintervals. The data of the hydrothermal system are taken from Test System 1 while wind data are taken from [45] and also reported in Table A3 in the Appendix A.

The implemented methods are coded on MATLAB and a personal computer with the CPU of Intel Core i7-2.4GHz, RAM 4GB for obtaining 50 successful runs. The optimal generations of two systems are reported in Tables A4 and A5 in the Appendix A.

5.1. Comparison Results on Test System 1

In this section, the MASCSA is tested on a large hydrothermal system with four hydropower plants and four thermal power plants, considering valve effects scheduled in twenty-four one-hour subintervals. In order to investigate the effectiveness of the MASCSA, CSA and SDCSA are implemented to compare the results. In the first simulation, P_s and $Iter^{max}$ are set to 200 and 5000 for all methods, respectively, but CSA cannot reach successful runs for each of the 50 trial runs. Meanwhile, SDCSA reaches a very low success rate. Then, $Iter^{max}$ is increased to 10,000 with a change of 1000 iterations. SDCSA and MASCSA can reach 100% successful runs at $Iter^{max} = 10,000$, but CSA only reaches 50 successful runs over 70 trial runs. Results obtained by the implemented methods are summarized in Table 1.

It is noted that the results from CSA, SDCSA, and MASCSA are obtained at $P_s = 200$ and $Iter^{max} = 10,000$, with the aim of reaching a higher number of successful runs for CSA and SDCSA. In order to check the powerful searchability of MASCSA over CSA and SDCSA, Figures 4 and 5 are plotted to present less cost and the corresponding level of improvement. Figure 4 indicates that the reduced cost that ASCSA can reach is significant and much increased for average cost and maximum cost. Accordingly, the level of improvement of the minimum cost, average cost, and maximum cost are respectively 0.54%, 1.3% and 2.81% as compared to CSA and 0.29%, 0.92% and 2.75% as compared to SDCSA. Similarly, the improvement of standard deviation is also high, corresponding to 23% and 27.12%, as compared to CSA and SDCSA. The indicated numbers lead to the conclusion that MASCSA is superior over CSA and SDCSA, in terms of finding the best solution and reaching a more stable search process.

In addition, the best run and the average run of 50 successful runs are also plotted in Figures 6 and 7 for search speed comparison. The two figures confirm that MASCSA is much faster than CSA and SDCSA for the best run and the average of all runs. In fact, in Figure 6, the best solution of MASCSA at the 5000th iteration is much better than CSA and SDCSA, and the best solution of MASCSA at the 7000th iteration is also better than that of CSA and SDCSA at the last iteration. This indicates that the speed of MASCSA can be nearly two times faster than CSA and SDCSA. In Figure 7, the average solution of 50 solutions found by MASCSA is also much more effective than that of CSA and SDCSA. The average solution of MASCSA at the 7000th iteration is also better than that of CSA and SDCSA at the last iteration. Clearly, the stability of MASCSA is also nearly twice as good as that of CSA and SDCSA. The whole view of the 50 solutions comparison can be seen by checking Figure 8. Many solutions of MASCSA have lower cost than that of CSA and SDCSA.

In summary, the proposed MASCSA is superior over CSA and SDCSA in finding optimal solutions and reaching a faster search speed for Test System 1. Hence, the proposed modifications of MASCSA are effective for large-scale power systems.

Table 1. Summary of results obtained by CSA, SDCSA, and MASCSA for Test System 1.

Method	CSA	SDCSA	MASCSA
Minimum Cost ($)	35640.09	35550.06	35447.25
Average Cost ($)	36835.21	36694.27	36355.55
Maximum Cost ($)	38616.82	38595.07	37533.4
Std. Dev. ($)	595.36	628.65	458.1301
Computation Time (s)	437.30	498.71	457.92
Success Rate	50/70	50/50	50/50

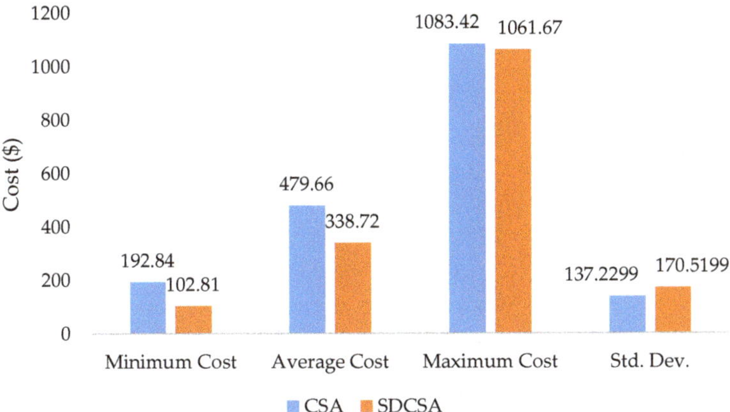

Figure 4. Better cost in $ obtained by MASCSA, compared to CSA and SDCSA, for Test System 1.

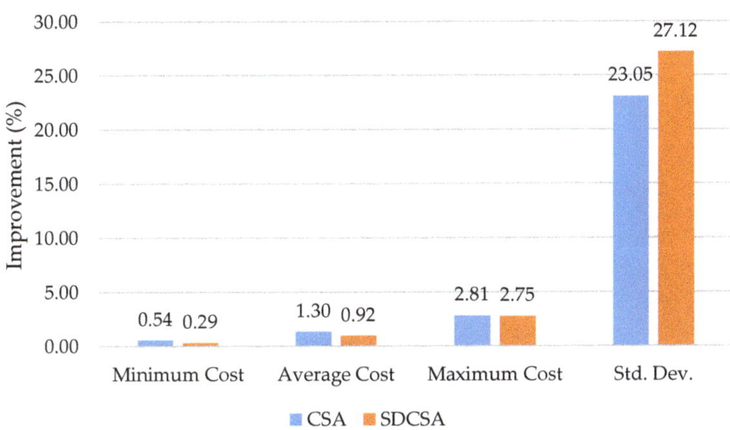

Figure 5. The level of improvement of MASCSA compared with CSA and SDCSA for Test System 1.

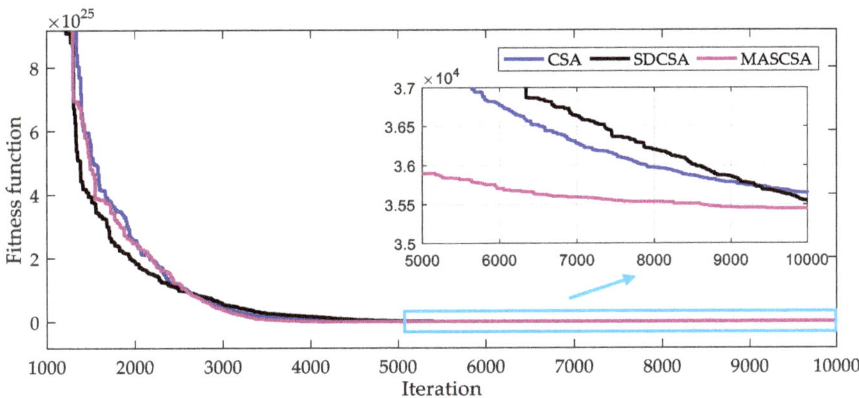

Figure 6. The best convergence characteristics obtained by implemented CSA methods for Test System 1.

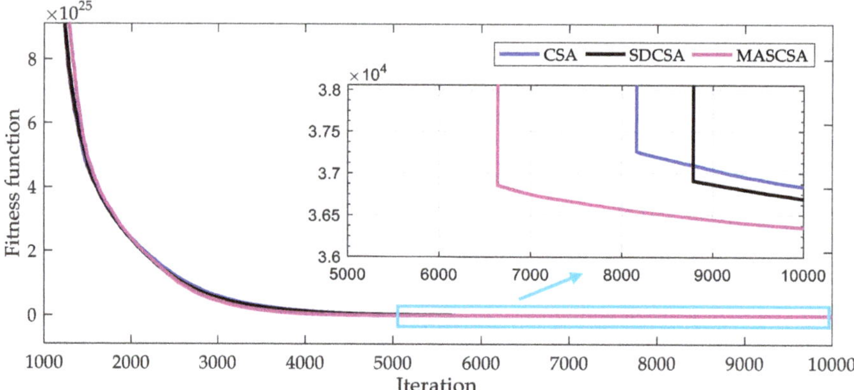

Figure 7. The mean convergence characteristics of 50 successful runs obtained by implemented CSA methods for Test System 1.

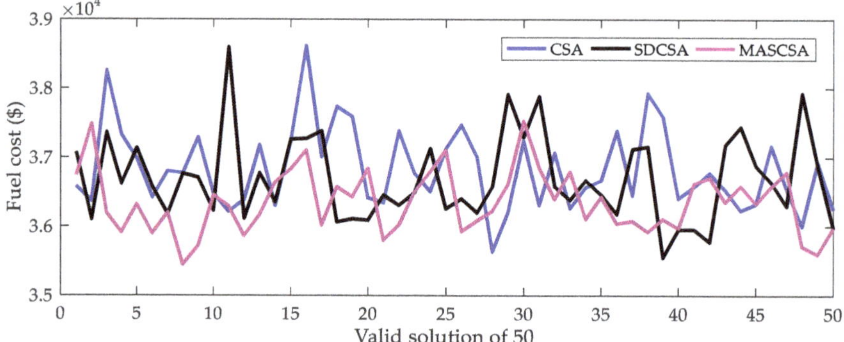

Figure 8. Fitness functions of 50 successful runs obtained by CSA methods for Test System 1.

5.2. Comparison Results on Test System 2

In this section, the implemented methods are tested on a wind-hydro-thermal system. The system is the combination of the hydrothermal system in Test System 1 and two wind farms. The system is optimally scheduled in twenty-four one-hour subintervals. Similar to Test System 1, three CSA methods, including CSA, SDCSA, and MASCSA, are successfully implemented considering all constraints of the system with the initial settings of $P_s = 200$ and $Iter^{max} = 10,000$. Accordingly, Table 2 shows the obtained results by CSA, SDCSA, and MASCSA. The key information in this table is the success rate comparison. Meanwhile, the comparison of cost is shown in Figures 9 and 10 for reporting less cost and the corresponding level of improvement of MASCSA over CSA and SDCSA, respectively. It should be emphasized that MASCSA can reach 50 successful runs over 50 trial runs, but the number of trial runs for CSA and SDCSA is much higher, which is 72 runs for CSA and 65 runs for SDCSA. Obviously, the constraint solving performance of MASCSA is much better than CSA and SDCSA. Figure 9 shows the significant cost reduction that MASCSA can reach as compared to CSA and SDCSA. The exact calculation, as compared to CSA and SDCSA, of MASCSA can reduce minimum cost by $685.51 and $422.90, mean cost by $572.95 and $466.75, maximum cost by $447.48 and $291.97, and standard deviation by 49.53 and 72.62. As can be observed from Figure 10, the level of improvement is also high and can be up to 2.46% for minimum cost and 14.69% for standard deviation.

Table 2. Summary of results obtained by CSA, SDCSA, and MASCSA for Test System 2.

Method	CSA	SDCSA	MASCSA
Minimum Cost ($)	27890.67	27628.06	27205.16
Average Cost ($)	28682.37	28576.17	28109.42
Maximum Cost ($)	29793.52	29638.01	29346.04
Std. Dev.	471.41	494.50	421.88
Computation Time (s)	440.5	499. 1	462.4
Success Rate (%)	50/72	50/65	50/50

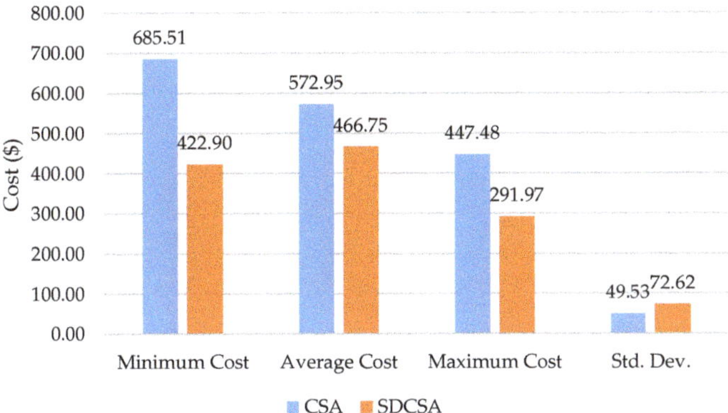

Figure 9. Better cost in $ obtained by MASCSA, compared to CSA and SDCSA for Test System 2.

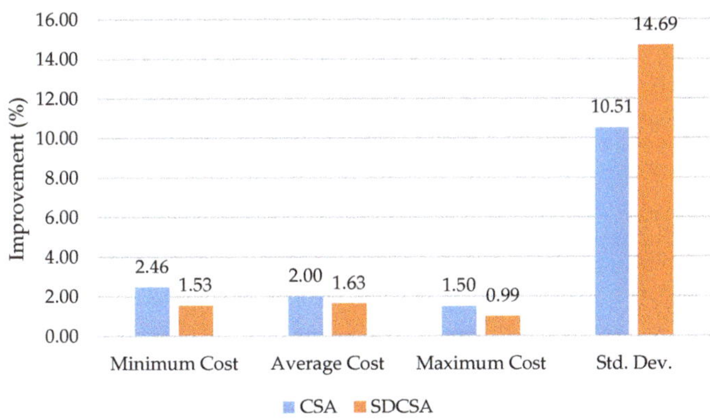

Figure 10. The level of improvement of MASCSA, compared to CSA and SDCSA for Test System 2.

Figures 11 and 12 illustrate the faster search performance of MASCSA than CSA and SDCSA for the best run and the whole search process of 50 successful runs. The pink curves of MASCSA in the two figures are always below the black and blue curves of CSA and SDCSA. The best solution and the mean solution of MASCSA are always more promising than those of CSA and SDCSA at each iteration. Namely, the best solution and the mean solution of MASCSA at the 7000th iteration have lower fitness functions than those of CSA and SDCSA at the 10,000th iteration. Fifty valid solutions shown in Figure 13 indicate that MASCSA can find a high number of better solutions than the best solution of CSA and SDCSA.

In summary, the proposed MASCSA can reach a higher success rate, better solutions, and faster speed than CSA and SDCSA for Test System 2. Consequently, the proposed MASCSA is really effective for the system.

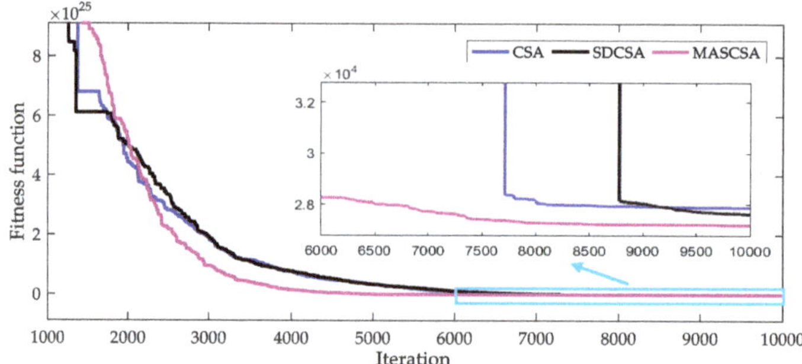

Figure 11. The best convergence characteristics obtained by implemented CSA methods for Test System 2.

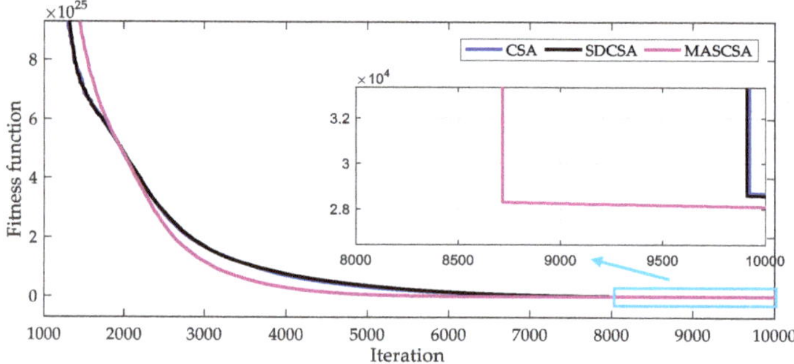

Figure 12. The mean convergence characteristics of 50 successful runs obtained by implemented CSA methods for Test System 2.

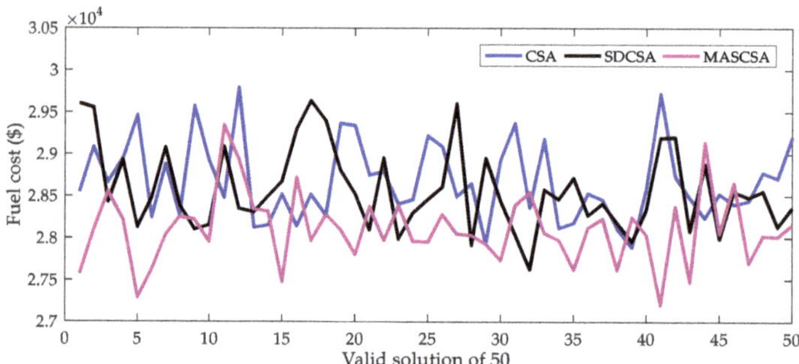

Figure 13. Fitness functions of 50 successful runs obtained by CSA methods for Test System 2.

6. Conclusions

In this paper, a Modified Adaptive Selection Cuckoo Search Algorithm (MASCSA) is implemented for determining the optimal operating parameters of a hydrothermal system and a wind-hydro-thermal system, to minimize the total electricity generation cost from all available thermal power plants. The fixed-head short-term model of hydropower plants is taken into consideration. All hydraulic constraints, such as initial and final reservoir volumes, the upper limit and lower limit of reservoir volume, and water balance of reservoir, are seriously considered. The proposed MASCSA competes with the conventional Cuckoo Search Algorithm (CSA) and Snap-Drift Cuckoo Search Algorithm (SDCSA). Two test systems are employed to run the proposed methods and those CSA methods.

The comparison results indicate that the proposed method is more powerful than CSA and SDCSA in searching for optimal solutions, with much faster convergence. The proposed method can deal with all constraints more successfully and reach much better results. The success rate of the proposed method is 100% for all test cases, while the success rates of the other CSA methods are 0% or much lower than 100%. Furthermore, the proposed method can reach a speed that is twice as fast as CSA and SDCSA. The improvement of the proposed method is significant compared to CSA methods, even when it is over 2%. Consequently, the proposed method is effective for complicated problems with a set of complicated constraints.

Author Contributions: T.T.N. and L.H.P. have simulated results and written the paper. L.C.K. has collected obtained results and analyzed results. F.M. was responsible for supervising, writing, and editing. All authors have read and agreed to the published version of the manuscript.

Funding: This research received no external funding.

Conflicts of Interest: The authors declare no conflicts of interest regarding the publication of this paper.

Nomenclature

TFC	Total fuel cost for generating electricity of all thermal power plants
t_i	Number of hours for the ith subinterval
$k_{tp}, m_{tp}, n_{tp}, \alpha_{tp}, \beta_{tp}$	Coefficients of the fuel cost function of the $t_p{}^{th}$ thermal power plant
$PT_{tp,i}$	Power generation of the $t_p{}^{th}$ thermal power plant at the ith subinterval
$PT_{tp,min}$	Minimum power generation of the $t_p{}^{th}$ thermal power plant
$PT_{tp,max}$	Maximum power generation of the $t_p{}^{th}$ thermal power plant
N_t	Number of thermal power plants
N_s	Number of subintervals
t_p	Thermal power plant index
h_p	Hydropower plant index
N_w	Number of wind turbines in a wind farm
N_h	Number of hydropower plants
$RV_{hp,i}$	Reservoir volume of the $h_p{}^{th}$ hydropower plant at the end of the ith subinterval
$WI_{hp,i}$	Water inflow into the reservoir of the $h_p{}^{th}$ hydropower plant at the ith subinterval
$Q_{hp,i}$	Total water discharge through turbines of the $h_p{}^{th}$ hydropower plant over the ith subinterval
$RV_{hp,start}$	Available reservoir volume of the $h_p{}^{th}$ hydropower plant before optimal scheduling
$RV_{hp,end}$	Final reservoir volume of the $h_p{}^{th}$ hydropower plant at the end of optimal scheduling
$RV_{hp,Ns}$	Reservoir volume of the $h_p{}^{th}$ hydropower plant at the end of the $N_s{}^{th}$ subinterval
$RV_{hp,min}$	Minimum reservoir volume of the $h_p{}^{th}$ hydropower plant
$RV_{hp,max}$	Maximum reservoir volume of the $h_p{}^{th}$ hydropower plant
$q_{hp,min}$	Minimum discharge per hour through turbines of the $h_p{}^{th}$ hydropower plant
$q_{hp,max}$	Maximum discharge per hour through turbines of the $h_p{}^{th}$ hydropower plant
$q_{hp,i}$	Discharge per hour through turbines of the $h_p{}^{th}$ hydropower plant over the ith subinterval
x_{hp}, y_{hp}, z_{hp}	Discharge function coefficients of the $h_p{}^{th}$ hydropower plant
$PH_{hp,min}$	Minimum power generation of the $h_p{}^{th}$ hydropower plant
$PH_{hp,max}$	Maximum power generation of the $h_p{}^{th}$ hydropower plant
$PT_{tp,max}$	Maximum power generation of the $t_p{}^{th}$ thermal power plant
w	Wind turbine index in the wind farm
$PW_{w,i}$	Power output of the wth wind turbine at the ith subinterval
N_w	Number of wind turbines in a wind farm
PW_w	Power generation of the wth wind turbine
$P_{TL,i}$	Total power loss at the ith subinterval
$P_{L,i}$	Power of load at the ith subinterval
$PW_{w,r}$	Rated generation of the wth turbine
WV_w	Wind speed flowing into the wind turbine
WV_{in}	Cut-in wind speed

WV_r	Rated wind speed
WV_{out}	Cut-out wind speed
$So_s{}^{new}$	The sth new solution
So_s	The sth solution
δ	Randomly generated number within 0 and 1
P_S	Population size
$Iter^{max}$	Maximum number of iterations
FF_s	Fitness function of the sth solution
$FF_s{}^{new}$	Fitness function of the sth new solution

Appendix A

Table A1. Data of thermal units for Test Systems 1 and 2.

Thermal Plant (tp)	k_{tp} ($/h)	m_{tp} ($/MWh)	n_{tp} ($/MW^{2}h)	α_{tp} ($/h)	β_{tp} (rad/MW)	$PT_{tp,min}$ (MW)	$PT_{tp,max}$ (MW)
1	60	1.8	0.0011	14	0.04	10	500
2	100	2.1	0.0012	16	0.038	10	675
3	120	1.7	0.0013	18	0.037	10	550
4	40	1.5	0.0014	20	0.035	10	500

Table A2. The data of hydropower plants of Test Systems 1 and 2.

Hydro Plant	x_{hp}	y_{hp}	z_{hp}	$PH_{hp,min}$ (MW)	$PH_{hp,max}$ (MW)	$RV_{hp,start}$ (acre-ft)	$RV_{hp,end}$ (acre-ft)	$RV_{hp,min}$ (acre-ft)	$RV_{hp,max}$ (acre-ft)
1	330	4.97	0.0001	0	1000	100,000	80,000	60,000	120,000
2	350	5.20	0.0001	0	1000	100,000	90,000	60,000	120,000
3	280	5.00	0.00011	0	1000	100,000	85,000	60,000	120,000
4	300	4.80	0.00011	0	1000	100,000	85,000	60,000	120,000

Table A3. Load demand and water inflows of Test Systems 1 and 2, and wind speed of Test System 2.

i	$P_{L,i}$ (MW)	$WI_{1,i}$ (acre-ft/h)	$WI_{2,i}$ (acre-ft/h)	$WI_{3,i}$ (acre-ft/h)	$WI_{4,i}$ (acre-ft/h)	$WV_{1,i}$ (m/s)	$WV_{2,i}$ (m/s)
1	1200	1000	800	800	600	13.2500	11.8000
2	1500	600	500	600	600	14.0000	12.0000
3	1100	700	500	700	700	12.7500	12.2000
4	1800	900	700	900	900	11.9000	12.4000
5	1200	900	700	900	900	12.5000	12.5000
6	1300	800	1000	800	800	13.9000	14.0000
7	1200	800	800	800	800	11.8000	15.0000
8	1500	700	800	700	700	12.7500	14.5000
9	1100	500	800	500	500	12.9000	13.0000
10	1800	500	800	500	500	12.2000	13.7500
11	1200	500	1000	500	500	15.0000	13.4000
12	1300	500	500	500	500	13.2500	13.4000
13	1200	800	500	700	800	14.3000	12.8000
14	1500	900	600	500	900	14.1000	12.2500
15	1100	600	600	600	600	14.2500	11.4000
16	1800	500	500	500	900	11.7500	11.5000
17	1200	950	950	950	900	13.7500	11.0000
18	1300	650	650	650	900	12.6000	11.2500
19	1200	550	550	550	700	11.5000	11.1000
20	1500	600	800	600	600	11.9000	11.0000
21	1100	600	800	600	600	14.5000	11.4500
22	1800	350	800	350	700	16.0000	11.8000
23	1200	600	1000	600	600	12.7000	11.7500
24	1300	400	400	800	800	13.0000	12.2500

Table A4. Optimal generations obtained by MASCSA for Test System 1.

i	$PH_{1,i}$ (MW)	$PH_{2,i}$ (MW)	$PH_{3,i}$ (MW)	$PH_{4,i}$ (MW)	$PT_{1,i}$ (MW)	$PT_{2,i}$ (MW)	$PT_{3,i}$ (MW)	$PT_{4,i}$ (MW)
1	44.80801	14.16502	490.1285	87.41643	89.3452	12.62116	88.98154	372.5341
2	609.7317	143.2493	12.82028	307.1734	23.53235	230.9884	137.2026	35.30191
3	109.6465	37.35409	2.452622	124.6044	28.81944	257.5346	262.3531	277.2352
4	118.6829	209.1928	666.0213	297.5565	12.51122	258.6675	166.9801	70.38769
5	56.98094	45.39613	206.5053	160.3263	78.1622	171.7098	110.8455	370.0738
6	503.2596	29.56263	129.0674	139.9014	22.80823	14.94842	269.5132	190.939
7	36.86074	25.08108	68.49895	347.292	94.01591	54.25231	265.6672	308.3319
8	354.087	119.3664	422.4064	22.18463	10.79384	40.37366	317.7927	212.9954
9	144.103	61.76661	44.56178	516.3614	25.04571	33.71163	15.01317	259.4367
10	655.2274	16.59057	91.88618	456.164	18.12036	176.5345	92.90249	292.5745
11	278.88	388.1928	55.58513	137.1538	91.58185	54.79326	122.1367	71.67637
12	139.7707	155.1218	691.5244	9.54121	32.3046	26.83425	147.7469	97.15622
13	303.2588	157.3504	313.9772	31.71765	83.83766	57.62889	109.2016	143.0277
14	10.34539	272.7907	410.6611	120.8436	18.08671	88.22264	305.0479	274.0019
15	88.28575	72.00485	91.75589	3.615984	125.1402	258.6483	264.4204	196.1286
16	202.7669	355.5148	124.7267	413.8378	21.4753	38.13277	185.3267	458.2191
17	405.6118	173.3598	65.81836	191.381	172.3582	81.61249	16.90782	92.95048
18	53.41923	578.2875	32.04454	36.24217	14.19755	206.3252	17.42234	362.0615
19	25.19439	55.43374	107.4736	606.1157	10.38145	17.65185	349.2747	28.47448
20	113.4698	387.7457	218.9941	476.7102	36.44216	61.98075	94.40057	110.2567
21	21.36867	45.56289	68.42235	0.01089	83.52566	176.4324	517.9546	186.7225
22	781.8121	182.5285	84.45162	206.7152	12.23156	10.00549	152.9166	369.3391
23	32.86834	39.96427	319.6729	24.81205	163.4837	28.45411	231.0542	359.6904
24	488.9518	0.799135	14.99782	405.5991	13.56384	100.0802	197.855	78.15313

Table A5. Optimal generations obtained by MASCSA for Test System 2.

i	$PH_{1,i}$ (MW)	$PH_{2,i}$ (MW)	$PH_{3,i}$ (MW)	$PH_{4,i}$ (MW)	$PT_{1,i}$ (MW)	$PT_{2,i}$ (MW)	$PT_{3,i}$ (MW)	$PT_{4,i}$ (MW)	$PW_{1,i}$ (MW)	$PW_{2,i}$ (MW)
1	16.85933	229.55	0.134587	148.8408	159.2501	36.35852	424.1347	31.472	99	54.4
2	109.9363	124.1835	450.167	129.3755	19.29588	30.2673	195.5862	277.1883	108	56
3	156.1532	89.33193	29.22211	424.3342	40.3847	18.70972	82.69941	108.5647	93	57.6
4	513.1342	271.1357	94.45502	420.9233	72.21183	249.6429	19.60898	16.88815	82.8	59.2
5	51.90915	130.3908	245.2019	379.6717	14.69276	115.6292	98.70234	13.80215	90	60
6	294.8739	11.9783	196.5268	34.98355	93.87338	35.32824	182.5234	271.1124	106.8	72
7	416.6924	85.92395	8.473279	179.9067	101.4721	22.10531	97.93109	125.8952	81.6	80
8	351.3595	155.4003	202.7161	240.5383	12.53609	48.06011	38.52946	281.8602	93	76
9	389.9609	30.64644	54.04259	10.0147	102.4636	86.01121	77.38582	190.6748	94.8	64
10	720.8998	144.4804	329.3808	181.5187	93.46645	33.37868	10.59862	129.8766	86.4	70
11	240.2481	189.7078	86.89813	349.9913	24.03108	13.15772	96.695	12.07097	120	67.2
12	244.4353	271.5002	395.8987	77.36695	41.96806	47.50929	45.08705	10.03442	99	67.2
13	168.0087	16.65194	56.36502	475.7266	11.10658	12.34142	73.02665	212.7731	111.6	62.4
14	388.4088	216.0905	196.185	6.14652	29.35371	72.20059	179.7867	244.6281	109.2	58
15	56.27707	87.74482	73.2046	34.60997	52.19771	161.4649	181.1577	291.1432	111	51.2
16	69.37554	645.5049	83.869	471.2085	10.57164	87.95043	180.7956	117.7243	81	52
17	24.40547	27.2146	408.8432	236.6929	27.55716	134.5309	146.7204	41.03535	105	48
18	402.7853	12.41216	333.9288	4.44101	33.34047	169.246	103.3128	99.33338	91.2	50
19	64.22907	202.1967	35.74582	90.85649	14.28666	246.1285	98.05939	321.6974	78	48.8
20	295.1399	75.00936	206.4338	254.5644	83.71423	100.8545	186.706	166.7778	82.8	48
21	36.08288	120.2875	402.702	24.58846	64.47947	139.7841	41.62437	104.8512	114	51.6
22	0.695207	18.65639	438.5363	708.044	58.3994	75.50154	141.1705	184.5967	120	54.4
23	178.9307	341.9207	198.9793	80.87282	22.25196	29.93504	10.96953	189.7399	92.4	54
24	396.5688	69.67454	216.289	157.7857	27.38508	154.2192	88.0156	36.06202	96	58

References

1. Nguyen, T.T.; Vu Quynh, N.; Duong, M.Q.; Van Dai, L. Modified differential evolution algorithm: A novel approach to optimize the operation of hydrothermal power systems while considering the different constraints and valve point loading effects. *Energies* **2018**, *11*, 540. [CrossRef]
2. Wood, A.; Wollenberg, B. *Power Generation, Operation and Control*; Wiley: New York, NY, USA, 1996.
3. Zaghlool, M.F.; Trutt, F.C. Efficient methods for optimal scheduling of fixed head hydrothermal power systems. *IEEE Trans. Power Syst.* **1988**, *3*, 24–30. [CrossRef]
4. Basu, M. Hopfield neural networks for optimal scheduling of fixed head hydrothermal power systems. *Electr. Power Syst. Res.* **2003**, *64*, 11–15. [CrossRef]
5. Wong, K.P.; Wong, Y.W. Short-term hydrothermal scheduling part. I. Simulated annealing approach. *IEE Proc. Gener. Transm. Distrib.* **1994**, *141*, 497–501. [CrossRef]
6. Yang, P.C.; Yang, H.T.; Huang, C.L. Scheduling short-term hydrothermal generation using evolutionary programming techniques. *IEE Proc. Gener. Transm. Distrib.* **1996**, *143*, 371–376. [CrossRef]
7. Hota, P.K.; Chakrabarti, R.; Chattopadhyay, P.K. Short-term hydrothermal scheduling through evolutionary programming technique. *Electr. Power Syst. Res.* **1999**, *52*, 189–196. [CrossRef]
8. Sharma, A.K.; Tyagi, R.; Singh, S.P. Short term hydrothermal scheduling using evolutionary programming. *Int. J. Inv. Res., Eng. Sci. and Tech.* **2014**, *1*, 19–23.
9. Chang, H.C.; Chen, P.H. Hydrothermal generation scheduling package: A genetic based approach. *IEE Proc. Gener. Transm. Distrib.* **1998**, *145*, 451–457. [CrossRef]
10. Sinha, N.; Chakrabarti, R.; Chattopadhyay, P.K. Fast evolutionary programming techniques for short-term hydrothermal scheduling. *IEEE Trans. Power Syst.* **2003**, *18*, 214–220. [CrossRef]
11. Nallasivan, C.; Suman, D.S.; Henry, J.; Ravichandran, S. A novel approach for short-term hydrothermal scheduling using hybrid technique. In Proceedings of the 2006 IEEE Power India Conference, New Delhi, India, 10–12 April 2006; p. 5.
12. Samudi, C.; Das, G.P.; Ojha, P.C.; Sreeni, T.S.; Cherian, S. Hydro thermal scheduling using particle swarm optimization. In Proceedings of the 2008 IEEE/PES Transmission and Distribution Conference and Exposition, Chicago, IL, USA, 21–24 April 2008; pp. 1–5.
13. Farhat, I.A.; El-Hawary, M.E. Short-term hydro-thermal scheduling using an improved bacterial foraging algorithm. In Proceedings of the 2009 IEEE Electrical Power & Energy Conference (EPEC), Montreal, QC, Canada, 22–23 October 2009; pp. 1–5.
14. Thakur, S.; Boonchay, C.; Ongsakul, W. Optimal hydrothermal generation scheduling using self-organizing hierarchical PSO. In Proceedings of the IEEE PES General Meeting, Minneapolis, MN, USA, 26 July 2010; pp. 1–6.
15. Türkay, B.; Mecitoğlu, F.; Baran, S. Application of a fast evolutionary algorithm to short-term hydro-thermal generation scheduling. *Energy Sources Part B Econ. Plan. Policy* **2011**, *6*, 395–405. [CrossRef]
16. Padmini, S.; Rajan, C.C.A. Improved PSO for Short Term Hydrothermal Scheduling. In Proceedings of the International Conference on Sustainable Energy and Interlligent Systems, Chennai, India, 20–22 July 2011; pp. 332–334.
17. Padmini, S.; Rajan, C.C.A.; Murthy, P. Application of improved PSO technique for short term hydrothermal generation scheduling of power system. In Proceedings of the International Conference on Swarm, Evolutionary, and Memetic Computing, Visakhapatnam, India, 19–21 December 2011; Springer: Berlin/Heidelberg, Germany, 2011; pp. 176–182.
18. Swain, R.K.; Barisal, A.K.; Hota, P.K.; Chakrabarti, R. Short-term hydrothermal scheduling using clonal selection algorithm. *Int. J. Electr. Power Energy Syst.* **2011**, *33*, 647–656. [CrossRef]
19. Fakhar, M.S.; Kashif, S.A.R.; Saqib, M.A.; ul Hassan, T. Non cascaded short-term hydro-thermal scheduling using fully-informed particle swarm optimization. *Int. J. Electr. Power Energy Syst.* **2015**, *73*, 983–990. [CrossRef]
20. Nguyen, T.T.; Vo, D.N.; Ongsakul, W. One rank cuckoo search algorithm for short-term hydrothermal scheduling with reservoir constraint. In Proceedings of the 2015 IEEE Eindhoven PowerTech, Eindhoven, The Netherlands, 29 June–2 July 2015; pp. 1–6.

21. Nguyen, T.T.; Vo, D.N.; Dinh, B.H. Cuckoo search algorithm using different distributions for short-term hydrothermal scheduling with reservoir volume constraint. *Int. J. Electr. Eng. Inform.* **2016**, *8*, 76. [CrossRef]
22. Dinh, B.H.; Nguyen, T.T.; Vo, D.N. Adaptive cuckoo search algorithm for short-term fixed-head hydrothermal scheduling problem with reservoir volume constraints. *Int. J. Grid Distrib. Comput.* **2016**, *9*, 191–204. [CrossRef]
23. Nguyen, T.T.; Vo, D.N.; Deveikis, T.; Rozanskiene, A. Improved cuckoo search algorithm for nonconvex hydrothermal scheduling with volume constraint. *Elektronika ir Elektrotechnika* **2017**, *23*, 68–73.
24. Nguyen, T.T.; Vo, D.N.; Dinh, B.H. An effectively adaptive selective cuckoo search algorithm for solving three complicated short-term hydrothermal scheduling problems. *Energy* **2018**, *155*, 930–956. [CrossRef]
25. Jadhav, H.T.; Roy, R. Gbest guided artificial bee colony algorithm for environmental/economic dispatch considering wind power. *Expert Syst. Appl.* **2013**, *40*, 6385–6399. [CrossRef]
26. Liu, X. Economic load dispatch constrained by wind power availability: A wait-and-see approach. *IEEE Trans. Smart Grid* **2010**, *1*, 347–355. [CrossRef]
27. Yuan, X.; Tian, H.; Yuan, Y.; Huang, Y.; Ikram, R.M. An extended NSGA-III for solution multi-objective hydro-thermal-wind scheduling considering wind power cost. *Energy Convers. Manag.* **2015**, *96*, 568–578. [CrossRef]
28. Zhou, J.; Lu, P.; Li, Y.; Wang, C.; Yuan, L.; Mo, L. Short-term hydro-thermal-wind complementary scheduling considering uncertainty of wind power using an enhanced multi-objective bee colony optimization algorithm. *Energy Convers. Manag.* **2016**, *123*, 116–129. [CrossRef]
29. Chen, Y.; Wei, W.; Liu, F.; Mei, S. Distributionally robust hydro-thermal-wind economic dispatch. *Appl. Energy* **2016**, *173*, 511–519. [CrossRef]
30. de Moraes, R.A.; Fernandes, T.S.; Arantes, A.G.; Unsihuay-Vila, C. Short-term scheduling of integrated power and spinning reserve of a wind-hydrothermal generation system with ac network security constraints. *J. Control. Autom. Electr. Syst.* **2018**, *29*, 1–14. [CrossRef]
31. Damodaran, S.K.; Sunil Kumar, T.K. Hydro-thermal-wind generation scheduling considering economic and environmental factors using heuristic algorithms. *Energies* **2018**, *11*, 353. [CrossRef]
32. He, Z.; Zhou, J.; Sun, N.; Jia, B.; Qin, H. Integrated scheduling of hydro, thermal and wind power with spinning reserve. *Energy Procedia* **2019**, *158*, 6302–6308. [CrossRef]
33. Cotia, B.P.; Borges, C.L.; Diniz, A.L. Optimization of wind power generation to minimize operation costs in the daily scheduling of hydrothermal systems. *Int. J. Electr. Power Energy Syst.* **2019**, *113*, 539–548. [CrossRef]
34. Daneshvar, M.; Mohammadi-Ivatloo, B.; Zare, K.; Asadi, S. Two-stage stochastic programming model for optimal scheduling of the wind-thermal-hydropower-pumped storage system considering the flexibility assessment. *Energy* **2020**, *193*, 116657. [CrossRef]
35. Dasgupta, K.; Roy, P.K.; Mukherjee, V. Power flow based hydro-thermal-wind scheduling of hybrid power system using sine cosine algorithm. *Electr. Power Syst. Res.* **2020**, *178*, 106018. [CrossRef]
36. Yang, X.S.; Deb, S. Cuckoo search via Lévy flights. In Proceedings of the 2009 World Congress on Nature & Biologically Inspired Computing (NaBIC) IEEE, Coimbatore, India, 9–11 December 2009; pp. 210–214.
37. Rakhshani, H.; Rahati, A. Snap-drift cuckoo search: A novel cuckoo search optimization algorithm. *Appl. Soft Comput.* **2017**, *52*, 771–794. [CrossRef]
38. Mohammadi, F.; Zheng, C. Stability Analysis of Electric Power System. In Proceedings of the 4th National Conference on Technology in Electrical and Computer Engineering, Tehran, Iran, 27 December 2018.
39. Mohammadi, F.; Nazri, G.-A.; Saif, M. A bidirectional power charging control strategy for plug-in hybrid electric vehicles. *Sustainability* **2019**, *11*, 4317. [CrossRef]
40. Mohammadi, F.; Nazri, G.-A.; Saif, M. An improved droop-based control strategy for MT-HVDC systems. *Electronics* **2020**, *9*, 87. [CrossRef]
41. Mohammadi, F.; Nazri, G.-A.; Saif, M. An improved mixed AC/DC power flow algorithm in hybrid AC/DC grids with MT-HVDC systems. *Appl. Sci.* **2020**, *10*, 297. [CrossRef]
42. Nguyen, T.T.; Mohammadi, F. Optimal placement of TCSC for congestion management and power loss reduction using multi-objective genetic algorithm. *Sustainability* **2020**, *12*, 2813. [CrossRef]
43. Shojaei, A.H.; Ghadimi, A.A.; Miveh, M.R.; Mohammadi, F.; Jurado, F. Multi-Objective Optimal Reactive Power Planning under Load Demand and Wind Power Generation Uncertainties Using ε-Constraint Method. *Appl. Sci.* **2020**, *10*, 2859. [CrossRef]

44. Yao, F.; Dong, Z.Y.; Meng, K.; Xu, Z.; Iu, H.H.C.; Wong, K.P. Quantum-inspired particle swarm optimization for power system operations considering wind power uncertainty and carbon tax in Australia. *IEEE Trans. Ind. Inform.* **2012**, *8*, 880–888. [CrossRef]
45. Zhang, H.; Yue, D.; Xie, X.; Dou, C.; Sun, F. Gradient decent based multi-objective cultural differential evolution for short-term hydrothermal optimal scheduling of economic emission with integrating wind power and photovoltaic power. *Energy* **2017**, *122*, 748–766. [CrossRef]

MDPI
St. Alban-Anlage 66
4052 Basel
Switzerland
Tel. +41 61 683 77 34
Fax +41 61 302 89 18
www.mdpi.com

Applied Sciences Editorial Office
E-mail: applsci@mdpi.com
www.mdpi.com/journal/applsci